Dimensioning, Tolerancing, and Gaging Applied

ref
ASME Y14.5M-1994
ANSI B4.4(B89.3.6)
and ISO 1101

Gary P. Gooldy

Prentice Hall
Upper Saddle River, New Jersey   *Columbus, Ohio*

Library of Congress Cataloging-in-Publication Data
    Dimensioning, tolerancing, and gaging applied / Gary P. Gooldy.
       p.   cm.
    Includes index.
    ISBN 0-13-791600-0 (pbk.)
    1. Engineering drawings--Dimensioning.   2. Tolerance (Engineering)
3. Mensuration.   I. Title.
T357.G65 1999
604.2'43   97-44661
                                                                  CIP

Cover art: Gary P. Gooldy
Editor: Stephen Helba
Production Editor: Patricia S. Kelly
Design Coordinator: Karrie M. Converse
Cover Designer: Rod Harris
Production Manager: Deidra M. Schwartz
Marketing Manager: Frank Mortimer, Jr.

This book was set in Arial by Gary P. Gooldy and was printed and bound by Banta Company. The cover was printed by Phoenix Color Corp.

 © 1999 by Prentice-Hall, Inc.
Simon & Schuster/A Viacom Company
Upper Saddle River, New Jersey 07458

Printed in the United States of America

10 9 8 7 6 5 4 3 2 1

ISBN: 0-13-791600-0

Prentice-Hall International (UK) Limited, *London*
Prentice-Hall of Australia Pty. Limited, *Sydney*
Prentice-Hall of Canada, Inc., *Toronto*
Prentice-Hall Hispanoamericana, S. A., *Mexico*
Prentice-Hall of India Private Limited, *New Delhi*
Prentice-Hall of Japan, Inc., *Tokyo*
Simon & Schuster Asia Pte. Ltd., *Singapore*
Editora Prentice-Hall do Brasil, Ltda., *Rio de Janeiro*

Preface

The purposes of *Dimensioning, Tolerancing and Gaging Applied* are to serve as supplemental material to National Standards ASME Y14.5M on dimensioning and tolerancing and ANSI B4.4 on inspection of workpieces, to explore the relationship of Y14.5M and B4.4 to the International Standards Organization (ISO) 9000 series quality standards, and to be used as a training aid. This workbook is primarily for those with basic or limited knowledge of the above subject, and is developed, organized and simplified to suit the average user. Other purposes are to share the author's experiences in the application of the principles of the ASME Y14.5M-1994 standard (and earlier standards) and provide gaging and measurement reference from ANSI B4.4 measurement standard to close the loop between design, manufacturing, and quality (inspection). Considerable figures are included for measurement and gaging reference. Engineers, drafters, designers, quality engineers, manufacturing engineers, inspectors, tool engineers, gage engineers, or persons from related fields who must prepare or take action on engineering drawings will benefit from this book.

Because this material is advisory, any issues unresolved should be directed to the national standard in question. Readers should have a background in basic math, blueprint reading, and drawing fundamentals. The figures in this book are complete only to the extent that they illustrate the point in question and may not reflect the best drafting/drawing practice.

The Y14.5 standard is considered an engineering specification language, but it contains information useful for manufacturing and quality control, providing a means for uniform interpretation and understanding between disciplines, as well as providing a bridge for a contract base, both nationally and internationally, for suppliers and customers.

The B4.4 standard is the current U.S. standard on Inspection of Workpieces, soon to be replaced by ASME B89.3.6, Gaging and Fixturing for GDT. Simple gage examples are used throughout the book.

Examples in this text contain figures with both metric and inch units, as the system is compatible with both. Exercises at the end of each chapter review the subject covered. At the end of the book are 11 general tests.

CERTIFICATION OF GDT PROFESSIONALS

A new standard, Y14.5.2, has been approved by the ASME. This standard sets forth the qualifications for two levels of certification. The first level, *Technologists*, provides a measure of individual ability to understand drawings prepared using the Y14.5 standard. The second level, *Senior*, provides a measure of individual ability to select and apply GDT controls of Y14.5 properly. This standard titled Certification of Geometric Dimensioning and Tolerancing Professionals (ASME Y14.5.2-1995) may be obtained through ASME.

TRAINING KITS

Material from this book is also available in instructor kits for presentation in workshops and seminars of one, two, or three days. This material may be used for full-semester programs. If you wish to purchase these kits with overhead transparencies and instructor guide notes, contact me at 812-372-9693.

I wish to express acknowledgment and appreciation to the ASME for approval to copy certain figures from ASME Y14.5M and B4.4 standards. Further, I am grateful to friends and colleagues who have helped in the development and improvement of the Y14.5 standard and therefore this work.

About the Author

Gary Gooldy has been involved with Y14.5 dimensioning since 1969 and has over 40 years' experience in U.S. industry in the areas of design, drafting, product engineering, training, and management. He is a member of the ANSI/ASME Y14.5M sub-committee on Dimensioning and Tolerancing and also ANSI/ASME B89.3.6 sub-committee on Gaging and Fixturing for GDT. He pioneered the development and use of Y14.5 for Cummins Engine Co. corporate engineering drawing practices. He has conducted GDT training seminars in the United States and Europe. His programs are approved by Society for Manufacturing Engineers and the American Society for Quality Control for recertification credits. He is president of GPG Consultants, Inc., serving as standards advisor and trainer for both industry and the academic community. Mr. Gooldy is retired from Cummins Engine Co., having served as manager of corporate drafting at the Cummins Technical Center.

Contents

v

1   INTRODUCTORY CONCEPTS

HISTORY

In the past, U.S. industry has not been aggressive in developing or adopting standard toleranceing practices. Large defense contractors have been required to conform and follow Military Standards (MIL-STDS) to acquire and keep military contracts. As evidenced by the country's spending on defense (as a percentage of Gross National Product), it is an economic necessity to do so, thus forcing the subcontract supplier base to follow as well. The marketplace is now a world stage, and the need to develop and build international markets to remain competitive reveals that, as a nation, we must think *global* in standards development and application. In addition, the U.S., Canada, and Mexico have established NAFTA. Countries in Europe have formed the European Committee for Standardization (CEN), the Asian community has Asian Standards Advisory Committee (ASAC), the U.S. and 11 other Latin American countries make up the Pan American Standards Commission (COPANT), while countries around the world develop and negotiate regional trade agreements.

Continued development and progress have led to increased commonality in worldwide standardization. The U.S. is a member of International Organization for Standards (ISO). All major countries of the world are also members. In addition to engineering standards, ISO has successfully directed the world's attention to universal quality standards, called ISO 9000. Further, the major economic countries of the world have individual country standards as follows:

United States	ANSI	United Kingdom	BSI
Japan	JIS	Italy	UNI
Germany	DIN	Canada	SCC
France	AFNOR	Australia	SAA

The ISO 9000 standards specify requirements for organizational quality systems. They do not specify the technical specifications of product, but do cover the systems that produce products. They are generic in that they apply to all products and industries. The ISO 9000 series is designed to compliment industry and/or product specific standards for quality to the customer relative to quality conformance.

Of the three standards, ISO 9001 has the broadest scope in that it includes design, production, inspection, testing, installation, and service elements. ISO 9000 is required in the European Norm (EN9000) standards of the European community; it has been adopted by NATO, and by no fewer than 80 of the world's major nations, giving assurances to intermarket quality. In the U.S., ANSI/ASQC QS9000 have been developed as the national standard equivalent.

ISO 9000 is not intended to be complicated, nor does it prescribe a specific method of product or process control. It does require that we *document* what we do, *act* as we have documented, and *produce* that level of quality for our customers. In doing so, we must determine what we want to document, what we are doing now, what the current quality is, what error rates are acceptable, and what elements must be brought up to that level. ISO requires that we be honest with ourselves and, above all, consistent.

This book attempts to help build a bridge from the conceptual world of ISO 9000 to the world of manufactured product. To do this, I believe we should start with engineering design and manufacturing and quality standards and link them through common elements. Starting from the design perspective of ASME Y14.5M-1994, we must be able to relate these principles and rules to the manufactured product, with appropriate quality controls to complete the process outlined in ISO 9000. We shall begin by reviewing general milestones in the U.S. standards development process.

Although it may be difficult to determine standards priority, the business agreement or marketplace will generally determine this. A summary of the evolution of U.S. standards is as follows:

1905	The *Taylor concept* introduces *limit gaging* for holes and shafts.
1929	*French's* manual for engineering drawings first mentions *drawing tolerancing.*
1935	The 18-page ASA Z14.1 American Standards Drawing and Drafting Room Practice which took ten years to develop is published.
1940	Chevrolet Draftsman's Handbook *introduces Maximum Material Condition (MMC)* to the U.S.
1941	Society of Automotive Engineers (SAE) Aircraft Engine Drafting Room Practice Manual introduces an *elementary form of true position* dimensioning.
1945	U.S. Army Ordnance Manual introduces *symbolism* for coaxiality, squareness, parallelism and true center position.
1949	MIL-STD-8 becomes the first military standard for dimensioning and tolerancing.
1953	MIL-STD-8A authorizes the use of *seven basic symbols* for GDT and introduces *functional dimensioning methodology.*
1957	American Standards Association (ASA) Y14.5 American Drawing Standard for Dimensioning and Notes is issued, containing 35 pages but *without symbols or tolerance* expression.
1959	MIL-STD-8B more closely aligns with ASA Y14.5 and SAE and accepts wider symbol usage. *Introduces three planes and limits of size* concepts, but is *not accepted* as an American Drawing Standard.
1963	MIL-STD-8C *expands true position* dimensioning and introduces *projected tolerance zones.*
1966	ASA becomes the United States of America Standards Institute (USASI) and issues USASI Y14.5 which introduced *symbols for cylindricity, runout, and profile, and changed the definition of straightness and concentricity. This was the first unified and accepted* American standard for dimensioning and tolerancing.
1973	USASI becomes the American National Standards Institute (ANSI) and issues ANSI Y14.5, incorporating *all symbology, rate tolerancing, composite positional tolerancing, introducing Least Material Condition (LMC), and datum targets, and allowing dual dimensioning methods.*
1982	ANSI Y14.5M-1982 undergoes sweeping change, including *symbols for counterbore, countersink, depth, square, total runout, conical taper, dimension origin, all-around profile and slope. Dual dimensioning is dropped, tolerancing for ISO limits and fits is introduced, 90 degree angles are implied, metric dimensions are covered, zero tolerancing at MMC is explained, and provision for Computer Aided Design (CAD) is introduced.*
1994	ANSI/ASME Y14.5M and Y14.5.1M issued.

SYMBOLS

The chart of Figure 1-1 shows graphically the development of symbols, in the above standards, starting in the 1940s and progressing thru ANSI Y14.5M-1982. Some key points to consider from figure 1-1 include:

The use of Concentricity and TIR simultaneously through 1966. The addition of Position symbol in MIL-STD 8A. The evolution of Straightness and Flatness. The changes in Feature Control Frames. The development and introduction of Runout 1966. The addition of Profile tolerancing 1966.The development of Symmetry 1945 until it was dropped in 1982, and resurrected 1994. The addition of Composite Position in 1973. The addition of supplemental symbols in 1982.

Symbolism continues to improve and much progress is being made in gaining agreement with ISO on standardized application of these symbols. See Figures 1-2 through 1-6.

TOLERANCE CHARACTERISTICS SYMBOL HISTORY

CHARACTERISTIC	D-30-1-7	MIL-STD-8A	MIL-STD-8B		MIL-STD-8C		USASI Y14.5-1966		ANSI Y14.5-1973		ANSI Y14.5M-1982	
			IF RFS	IF MMC	IF RFS	IF MMC	IF RFS	IF MMC	IF RFS	IF MMC	IF RFS	IF MMC
CONCENTRICITY (COAXIALITY)	◉ A XXX	◎ A XXX TIR	◎ A XXX TIR	◎ AⓂ XXXTIRⓂ	◎ A XXX TIR	◎ AⓂ XXXTIRⓂ	↗ A XXX ◎ A XXX DIA	⊕ A XXX DIA	↗ A XXX TOTAL ◎ A ⌀XXX	⊕ AⓂ ⌀XXXⓂ	↗↗ XXX A ◎ ⌀XXX A	⊕ ⌀XXXⓂ AⓂ
PERPENDICULARITY (SQUARENESS)	⊥ A XXX	⊥ A XXX	⊥ A XXX	⊥ AⓂ XXXⓂ	⊥ A XXX	⊥ AⓂ XXXⓂ	⊥ A XXX	⊥ AⓂ XXXⓂ	⊥ A XXX	⊥ AⓂ XXXⓂ	⊥ XXX A	⊥ XXXⓂ AⓂ
PARALLELISM	∥ A XXX	∥ A XXX	∥ A XXX	∥ AⓂ XXXⓂ	∥ A XXX	∥ AⓂ XXXⓂ	∥ A XXX	∥ AⓂ XXXⓂ	// A XXX	// AⓂ XXXⓂ	// XXX A	// XXXⓂ AⓂ
SYMMETRY	ȼ A XXX		≡ A XXX	≡ AⓂ XXXⓂ	≡ A XXX	≡ AⓂ XXXⓂ	≡ A XXX TOTAL ⊕ AⓈ XXX TOTALⓈ	⊕ A XXX TOTAL	≡ A XXX ⊕ AⓈ XXXⓈ	⊕ AⓂ XXXⓂ	⊕ ⌀XXXⓈ AⓈ	// AⓂ XXXⓂ
ANGULARITY		∠ A XXX	∠ A XXX	∠ AⓂ XXXⓂ	∠ A XXX	∠ AⓂ XXXⓂ	∠ A XXX	∠ AⓂ XXXⓂ	∠ A XXX	∠ AⓂ XXXⓂ	∠ A XXX	∠ XXXⓂ AⓂ
INTERRELATED SURFACE CHARACTERISTICS	◉ A ⊥ D XXX		◎ A ⊥ D XXX TIR (may apply feature	◎ AⓂ ⊥ D XXXTIRⓂ RFS - datum MMC)	◎ A ⊥ D XXX TIR (may apply feature	◎ AⓂ ⊥ D XXX TIRⓂ RFS - datum MMC)	↗ A B XXX (may apply feature	⊕ A D XXX DIA RFS - datum MMC)	↗ A B XXX TOTAL (may apply feature	⊕ AⓂ B ⌀XXXⓂ RFS - datum MMC)	↗↗ XXX A B (may apply feature	⊕ ⌀XXXⓂ AⓂ B RFS - datum MMC)
FLATNESS		∼ XXX	⌒ XXX		— XXX		▱ XXX		▱ XXX		▱ XXX ▱ XXX .010/1.00 X 1.00	
STRAIGHTNESS		∼ XXX	∼ XXX	∼ XXXⓂ	— XXX	— XXX Ⓜ	— XXX		— XXX — ⌀XXX	— ⌀XXXⓂ	— XXX — ⌀XXX — ⌀XXX ⌀.010/1.00 X 1.00	— ⌀XXXⓂ
POSITION (HOLES)		⊕ A XXX DIA	⊕ AⓈ XXX DIAⓈ	⊕ A XXX DIA	⊕ AⓈ XXX DIAⓈ	⊕ A XXX DIA	⊕ AⓈ XXX DIAⓈ	⊕ A XXX DIA	⊕ AⓈ ⌀ XXX Ⓢ	⊕ AⓈ ⌀ XXX Ⓢ	⊕ ⌀XXXⓈ AⓈ	⊕ ⌀XXXⓂ A Ⓜ
POSITION (SLOTS)			⊕ AⓈ XXX TOTALⓈ	⊕ A XXX TOTAL	⊕ AⓈ XXX TOTALⓈ	⊕ A XXX TOTAL	⊕ AⓈ XXX TOTALⓈ	⊕ A XXX TOTAL	⊕ AⓈ XXXⓈ	⊕ AⓂ XXX Ⓜ	⊕ XXX Ⓢ A Ⓢ	⊕ XXX Ⓜ A Ⓜ
CIRCULARITY (ROUNDNESS)			○ XXX (DIAMETRAL)		○ XXX ON R ○ XXX (DIAMETRAL)		○ XXX		○ XXX		○ XXX	
PROJECTED TOL ZONE					⊕ XXX DIA Ⓢ TOL ZONE PROJ XXX	⊕ XXX DIA TOL ZONE PROJ XXX	⊕ XXX DIA Ⓢ TOL ZONE PROJ XXX	⊕ XXX DIA TOL ZONE PROJ XXX	⊕ A ⌀ XXX Ⓢ XXX Ⓟ	⊕ A ⌀ XXX Ⓜ XXX Ⓟ	⊕ ⌀ XXX Ⓢ A XXX Ⓟ	⊕ ⌀ XXX Ⓜ A XXX Ⓟ
PROFILE - LINE							⌒ XXX		⌒ XXX		⌒ XXX	
PROFILE - SURFACE						USED NOTE	⌓ XXX		⌓ A XXX		⌓ A XXX	
CIRCULAR RUNOUT							↗ A XXX CIRCULAR		↗ A XXX		↗ XXX A	
TOTAL RUNOUT							↗ A XXX		↗ A XXX TOTAL		↗↗ XXX A	
CYLINDRICITY							⌭ XXX		⌭ XXX		⌭ XXX	
COMPOSITE POSITION CONTROL									⊕ A B C ⌀XXXⓈ A ⌀XXX Ⓢ	⊕ A B C ⌀XXXⓂ A ⌀XXX Ⓜ	⊕ ⌀XXXⓈ A B C ⌀XXXⓈ A	⊕ ⌀XXXⓂ A B C ⌀XXXⓂ A
RUNOUT COAXIAL DATUMS									↗ A - B XXX TOTAL		↗↗ XXX A - B	

NEW OR REVISED BY ASME Y14.5M-1994

SYMMETRY ≡ TANGENT PLANE Ⓣ FREE STATE Ⓕ STATISTICAL TOL ⓈⓉ

CONTROLLED RADIUS CR BETWEEN ↔ RFS none DATUM FEATURE ▲[A]

Figure 1-1 The evolution of symbols in the United States, 1945-1982.

SYMBOL FOR:	ASME Y14.5M	ISO
STRAIGHTNESS	—	—
FLATNESS	▱	▱
CIRCULARITY	○	○
CYLINDRICITY	⌭	⌭
PROFILE OF A LINE	⌒	⌒
PROFILE OF A SURFACE	⌓	⌓
ALL AROUND	 ⌲	 ⌲ (proposed)
ANGULARITY	∠	∠
PERPENDICULARITY	⊥	⊥
PARALLELISM	//	//
POSITION	⌖	⌖
CONCENTRICITY (concentricity and coaxiality in ISO)	◎	◎
SYMMETRY	⌯	⌯
CIRCULAR RUNOUT	*↗	↗
TOTAL RUNOUT	*⌰	⌰
AT MAXIMUM MATERIAL CONDITION	Ⓜ	Ⓜ
AT LEAST MATERIAL CONDITION	Ⓛ	Ⓛ
REGARDLESS OF FEATURE SIZE	NONE	NONE
PROJECTED TOLERANCE ZONE	Ⓟ	Ⓟ
TANGENT PLANE	Ⓣ	Ⓣ (proposed)
FREE STATE	Ⓕ	Ⓕ
DIAMETER	⌀	⌀
BASIC DIMENSION (theoretically exact dimension in ISO)	50	50
REFERENCE DIMENSION (auxiliary dimension in ISO)	(50)	(50)
DATUM FEATURE	 *⊐Ⓐ	 ▲ or  *⊐Ⓐ

* MAY BE FILLED OR NOT FILLED

Figure 1-2 Comparison of symbols.
(copied with permission from ASME Y14.5M-1994)

4

SYMBOL FOR:	ASME Y14.5M	ISO
DIMENSION ORIGIN	⬦—▶	⬦—▶
FEATURE CONTROL FRAME	⊕ ⌀0.5Ⓜ A B C	⊕ ⌀0.5Ⓜ A B C
CONICAL TAPER	▷	▷
SLOPE	◁	◁
COUNTERBORE/SPOTFACE	⊔	⊔ (proposed)
COUNTERSINK	∨	∨ (proposed)
DEPTH/DEEP	↧	↧ (proposed)
SQUARE	□	□
DIMENSION NOT TO SCALE	15	15
NUMBER OF PLACES	8X	8X
ARC LENGTH	⌒105	⌒105
RADIUS	R	R
SPHERICAL RADIUS	SR	SR
SPHERICAL DIAMETER	S⌀	S⌀
CONTROLLED RADIUS	CR	NONE
BETWEEN	*◀—▶	NONE
STATISTICAL TOLERANCE	⟨ST⟩	NONE
DATUM TARGET	⌀6/A1 or (A1)—⌀6	⌀6/A1 or (A1)—⌀6
TARGET POINT	✕	✕

* MAY BE FILLED OR NOT FILLED

**Figure 1-3 Supplementary symbols.
(copied with permission from ASME Y14.5M-1994)**

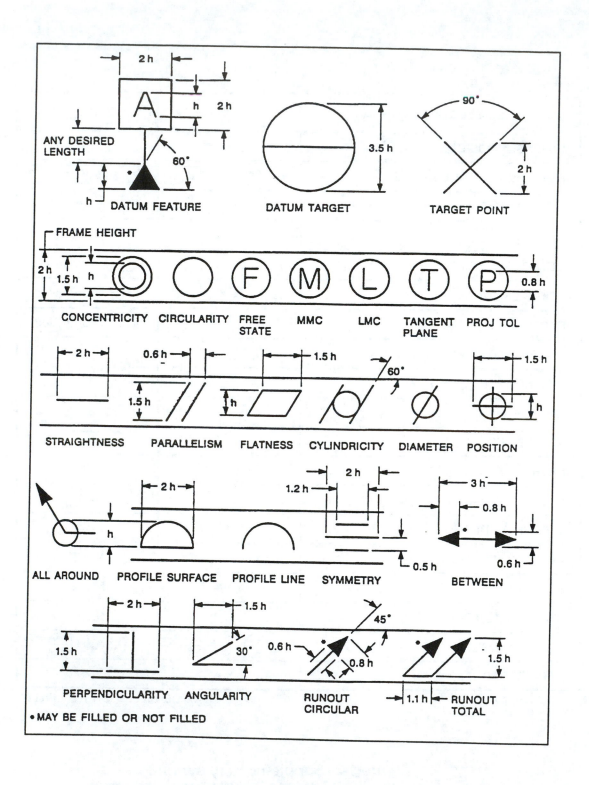

Figure 1-4 Form and proportion: geometric controls symbols.
(copied with permission from ASME Y14.5M-1994)

6

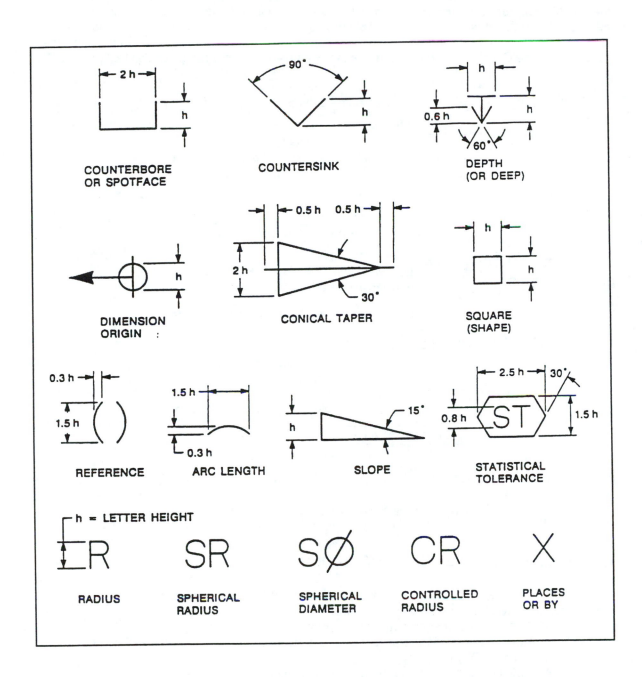

Figure 1-5 Form and proportion (continued).
(copied with permission from ASME Y14.5M-1994)

FEATURE CONTROL FRAMES

Now that we are familiar with symbols, we may arrange them to form symbol sentences by the use of *Feature Control Frames*. The symbols are placed in a specific order within the control boxes, with the geometric tolerance control first, the tolerance zone shape and tolerance value second, and the datum(s) reference third. See Figure 1-6 and ASME Y14.5M-1994 for additional data on Feature Control Frames. These *Control Frames* are used throughout the text.

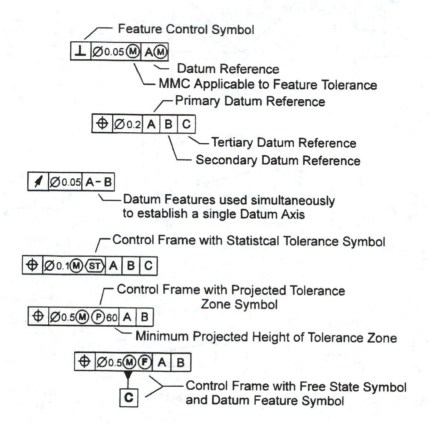

Figure 1-6 Feature Control Frames.

FUNDAMENTAL RULES OF DIMENSIONING

We should be familiar with some basic drawing and quality rules. The rules shown in on page 9 have been extracted from the following standards:

ASME Y14.5M-1994 Dimensioning and Tolerancing
ANSI B4.4 Inspection of Workpieces (soon to be replaced by ASME B89.3.6 Gaging and Fixturing for GDT)
ANSI B89.6.2 Temperature and Environmental Controls
ASME Y14.2 Line Conventions and Lettering

These *Fundamental Rules* should be clearly understood; with the *General Rules* of Figures 1-15 thru 1-18, they form the foundation of the design, documentation, production processes discussed earlier, and as outlined in ISO 9000.

FUNDAMENTAL RULES OF DIMENSIONING

Extracted from Y14.5 Y14.2 B89.3.6 B4.4 B89.6.2

1. Dimensions shall be toleranced. (Exceptions: reference, minimum, maximum, standard stock.)
2. Dimensions shall be complete with no more dimensions than necessary, and with no *double dimensions*.
3. Dimensions not to scale shall be *underlined.*
4. Basic dimensions require an associated tolerance control.
5. Drawings shall define *functional requirements* without specifying manufacturing methods, unless those methods are critical to precise *engineering functional requirements*.
6. Nonmandatory processing dimensions and/or information may be noted as [MFG DATA].
7. Dimension lines shown crossing at 90 degrees are implied to apply at 90 degrees.
8. Dimensions and tolerances apply at 20 degrees C, with humidity level at 45% maximum.
9. **Dimensions and tolerances apply in the *free state* unless noted.**
10. Dimension and tolerance application shall be clear on *coated or treated part surfaces.*
11. Decimal dimensions shall be used.
12. Dimensions should be shown in profile views, to visible outlines.
13. Dimensions should appear *outside* object lines.
14. **The length, width, or height (depth) of the *tolerance zone* is equal to the length, width, or height of the *feature*, unless shown otherwise.**
15. **Dimensions and tolerances apply only at the drawing level shown and are not mandatory at any higher level (assembly) that may be altered by process or assembly stresses.**
16. To ensure proper understanding of drawing requirements, and for future reference, the drawing should specify the *standard of reference origin,* such as ASME Y14.5M-1994.

DEFINITIONS

The definitions that follow are similar to those of ASME Y14.5, with some modifications for simplicity. These definitions are also part of the *foundation* discussed earlier and are terms used throughout the Y14.5 standard. They should be thoroughly understood and followed for consistency and continuity. ASME Y14.5 is the authority for precise wording. Other brief descriptions are found in the Glossary .

Actual Local Size: The measured value of any opposed elements, or cross section of a feature.

Actual Mating Size (Envelope): The value of the actual mating envelope (the smallest possible cylinder about an external feature or the largest possible cylinder within an internal feature).

Actual Size:The general term for the size of a feature including the actual mating size and actual local size.

Basic Dimension: A numerical value that describes the exact theoretical size, shape, location, and other characteristics of a feature, datum, or datum target. The value is contained within a box.

Centerline: Generic term generally used when referring to an axis.

Centerplane: Generic term generally used when referring to the median plane of a feature or datum.

Datum: A theoretically exact point, line, or plane derived from a feature true counterpart.

Datum Feature: An actual part (workpiece) feature, including irregularities of flatness, circularity, cylindricity, and/or straightness.

Datum Target: A point, line, or area on a part feature used to establish a datum to ensure consistency and repeatability in processing and/or measurement.

Dimension: A numerical value to define part characteristics of size, geometry, and the like.

Feature: General term to describe a physical portion of a part. (i.e., a hole, surface, or slot).

Feature Axis: A straight line that coincides with the centerline of the feature counterpart.

Feature Centerplane: A true plane that coincides with the centerplane of a feature counterpart.

Feature of Size: A feature with opposed feature elements. (i.e., cylindrical, hexagonal, spherical, or two parallel surfaces).

Free State: The condition of a part absent of any restraining forces.

Full Indicator Movement (FIM): The total movement of the indicator device when measuring surface error.

Functional Gage: A term to describe a dedicated or limit gage that will receive the part with no force applied. Elements of the gage may be variable or movable.

Inner Boundary: The smallest boundary generated by the minimum feature size less the stated geometric tolerance and any additional bonus tolerance due to feature departure from its stated material condition.

Outer Boundary: The largest boundary generated by the maximum feature size plus the stated geometric tolerance and any additional bonus tolerance due to feature departure from its stated material condition.

Least Material Condition (LMC): The condition in which the feature contains the least material (i.e., when the part weighs the least).

Maximum Material Condition (MMC): The condition in which the feature contains the maximum material (i.e., when the part weighs the most).

Median Line: An imperfect line that passes through all the centerpoints of all cross sections of a feature, normal to the actual-size mating envelope.

Median Plane: An imperfect plane that passes through all the centerpoints of all cross sections of a feature, normal to the actual size mating envelope.

Nominal Size: Term for general identification.

Reference Dimension: A dimension without tolerance used for information only.

Regardless of Feature Size (RFS): The condition to indicate that a datum reference or geometric tolerance applies at any size increment within the size tolerance limits.

Resultant Condition: The boundary generated by the collective effects of size, material condition, geometric tolerance, and any bonus tolerances due to the feature departure from its stated material condition.

Simulated Datum: A datum simulated by processing or inspection equipment surfaces, surface plates, or tool centers (datum simulators).

Tangent Plane: A theoretically exact plane that contacts a feature surface at the high points on the surface.

Tolerance, Bilateral: A tolerance that exists in two directions from a specified dimension.

Tolerance, Bonus: An increase in form, orientation, or position tolerance allowed, equal to a feature's departure from the stated material condition size.

Tolerance, Geometric: General term for the categories of Form, Orientation, Profile, Location, and Runout tolerances.

Tolerance, Statistical: Assignment of tolerances related to formulas based on square-root sum of the tolerances.

Tolerance, Unilateral: A tolerance that exists in only one direction from a specified dimension.

True Geometric Counterpart: A theoretically perfect boundary (virtual condition mating envelope) of a feature.

True Position: The theoretically exact location of a feature established by basic dimensions.

Virtual Condition: A constant worst-case boundary generated by the collective effects of a size feature's specified material condition and the geometric tolerance for that material condition.

TOLERANCE

Dimensions must have limits. Limits determine the fits of mating parts, and Preferred Limits and Fits are covered by ANSI B4.1 (U.S. customary units) and B4.2 (metric units). Tolerance impacts both quality and cost, from the initial design concept through the manufacturing and service development process to the customer. Tolerance initially impacts new products in new tooling costs, process time, equipment maintenance, quality planning and audit, process control, scrap and salvage, and measurement capability or gaging costs. Hidden costs due to tolerancing inefficiencies or misunderstanding may be found in virtually every product if we look hard enough. To avoid these hidden costs, we should be able to justify and defend design tolerances both logically and mathematically.

DESIGN TOLERANCE DISTRIBUTION

Print tolerances are values used by various agencies in the manufacturing process. Tolerances also represent engineering requirements that must be met for product acceptance. For any given tolerance, the inspection process is allowed 10% of the print tolerance value. This process is covered in ANSI B4.4 (being replaced by B89.3.6) Inspection of Workpieces. Tooling manufacturers generally have 15% to 20% of the print tolerance for tooling, which in theory leaves manufacturing with 70% of the value for the process. This 70% is based on new equipment, and as tooling and fixtures wear, adjustments must be made. We rely on the process control system and line gaging or on audits to maintain conformance to the tolerances. Figure 1-7 graphically illustrates the breakdown of tolerance to achieve a typical quality level. This figure represents a standard CPK 1.33 quality level that is common in industry. Many industries are working toward much tighter limits as quality goals.

PREFERRED LIMITS AND FITS

ANSI B4.1 (U.S. customary units) covers:
RC running and sliding fits
LC location clearance fits
LT location transition fits
LN location interference fits
FN force fits

Each of these classes is subdivided into groups, including:

9 RC clearance classes of fits. The amount of tolerance is proportional to the size of the bore/shaft.
11 LC location classes of fits. These range from small clearance (LC1) to liberal clearance (LC11). LC fits provide assembly clearance, but are not intended for moving/mating durfaces.
6 LT classes of fits that may result in either a clearance or interference condition.
3 LN classes of fits for location of parts when a clearance is not desired.
5 FN classes of force or shrink fits. These provide constant pressure (press) throughout a range of sizes.

ANSI B4.2 (metric) covers preferred metric limits and fits, and is based on preferred sizes selected from ANSI B32.4. The first choice sizes are rounded from the Renard 10 (R10) series of preferred numbers. See Figures 1-8a through 1-8c.

12

Givens: From ANSI/ASME B89.3.6 (B4.4)
          Gaging tolerance is allowed 10% of
          feature size tolerance.

          Manufacturers Tooling Tolerance is generally
          considered as 15% to 20%of Feature Print
          Tolerance.

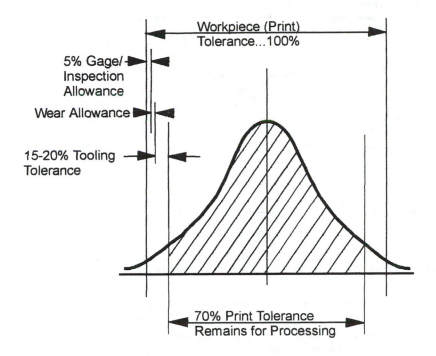

We might therefore conclude:
Manufacturing process capability on new equipment
should be equal to or better than 70% of drawing feature
tolerance, and also should have repeatability capability
of 99.7%, or +/- 3 sigma (1.33 CPK).

Figure 1-7 Normal tolerance distribution.

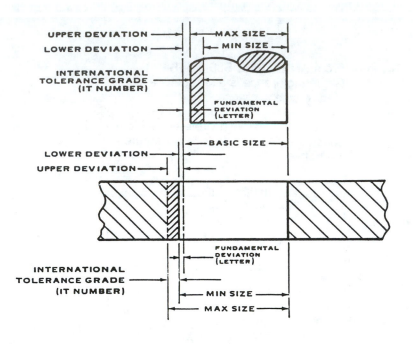

ISO SYMBOL		DESCRIPTION
Hole Basis	Shaft Basis	
H11/c11	C11/h11	*Loose running* fit for wide commercial tolerances or allowances on external members.
H9/d9	D9/h9	*Free running* fit not for use where accuracy is essential, but good for large temperature variations, high running speeds, or heavy journal pressures.
H8/f7	F8/h7	*Close running* fit for running on accurate machines and for accurate location at moderate speeds and journal pressures.
H7/g6	G7/h6	*Sliding* fit not intended to run freely, but to move and turn freely and locate accurately.
H7/h6	H7/h6	*Locational clearance* fit provides snug fit for locating stationary parts; but can be freely assembled and disassembled.
H7/k6	K7/h6	*Locational transition* fit for accurate location, a compromise between clearance and interference.
H7/n6	N7/h6	*Locational transition* fit for more accurate location where greater interference is permissible.
H7/p6	P7/h6	*Locational interference* fit for parts requiring rigidity and alignment with prime accuracy of location but without special bore pressure requirements.
H7/s6	S7/h6	*Medium drive* fit for ordinary steel parts or shrink fits on light sections, the tightest fit usable with cast iron.
H7/u6	U7/h6	*Force* fit suitable for parts which can be highly stressed or for shrink fits where the heavy pressing forces required are impractical.

(Left margin labels: Clearance Fits, Transition Fits, Interference Fits)
(Right margin labels: More Clearance, More Interference)

Figure 1-8 Preferred metric limits and fits.
(copied with permission from ASME Y14.5M-1994)

14

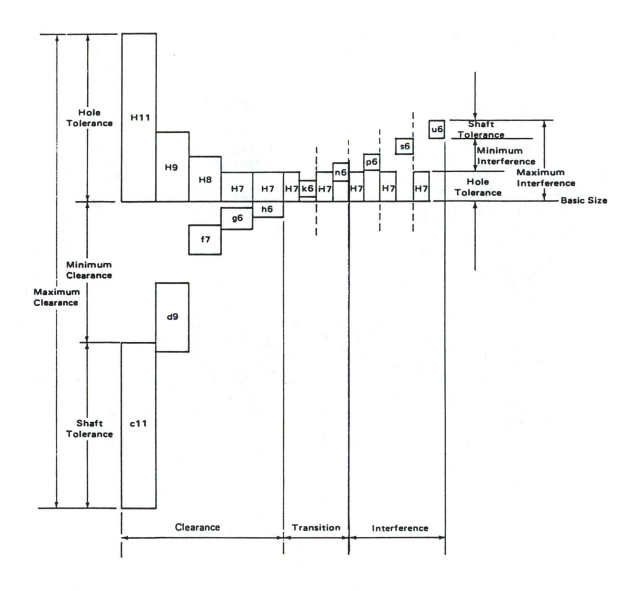

**Figure 1-8 continued: Hole basis fits.
(copied with permission from ASME B4.2)**

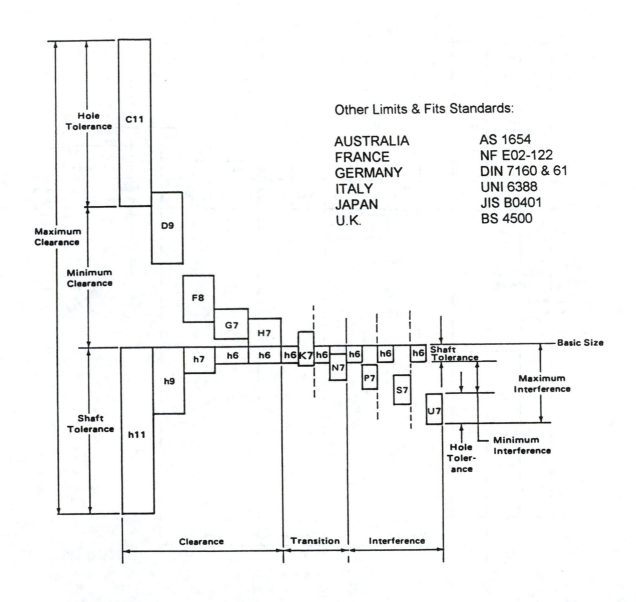

Other Limits & Fits Standards:

AUSTRALIA	AS 1654
FRANCE	NF E02-122
GERMANY	DIN 7160 & 61
ITALY	UNI 6388
JAPAN	JIS B0401
U.K.	BS 4500

Figure 1-8 continued: Shaft basis fits.
(copied with permission from ASME B4.2)

GAGING AND MEASUREMENT FUNDAMENTALS

Now that we have a picture of preferred limits and fits, fundamental principles can be applied to the quality and measurement process. The following have been extracted from ANSI B4.4, Inspection of Workpieces:

General: no part of gaging surface shall infringe on the MMC envelope.
Inspection and measurement tolerance is contained within the workpiece size limits.
Workpiece limits: interpretation of size limits.
    Holes: Largest perfect imaginary cylinder.
    Shafts: Smallest perfect imaginary cylinder.

FUNDAMENTAL GAGING PRINCIPLES [Ref B4.4 (B89.3.6) B89.1.9 B89.6]

1. Drawing dimensions and tolerances apply in the free state condition, and at 20 deg. C.
2. The recommended humidity level for measurement should not exceed 45%.
3. Gaging tolerance is normally considered as 5% of the *Feature Size Tolerance,* with an additional 5% for gage wear allowance.
4. The composite *Form Control* (form, orientation, profile and runout) of fixed gages is allowed 50% of gaging tolerance, and shall be contained within the Feature size tolerance values.
5. Gage block uncertainty is allowed 25% of Plug Gage tolerance.
6. Gage tolerances are normally additive to the gage element (material is added to the gage).
7. Gage Pins: Minimum Gage Pin feature length = feature thickness/depth.
   Pilot diameter should be approx. 30% larger than gage pin dia.
   Pilot diameter engagement length (with or without bushings) should be a minimum of 3X pilot diameter.
   Pins that pass thru riser blocks should have a pilot diameter engagement length of 4X pilot diameter, minimum.

8. Gagemakers Tolerance Classes:

Class	Roughness Average (Ra)	
	Micrometers	ISO 1302 No.
ZM	0.2	N4
YM	0.1	N3
XM	0.1	N3
XXM	0.05	N2
XXM	0.05	N2

The fundamental gage elements are shown in Figure 1-9.

The following standards provide additional information:

ANSI B47.1 Gage Blanks
ANSI B89.1.9 Precision Gage Blocks
ISO 463 Dial Gages
ISO 1302 Technical Drawings-Surface Texture
ISO 3650 Gauge Blocks
ASME Y14.5.1M Mathematical definitions for Y14.5

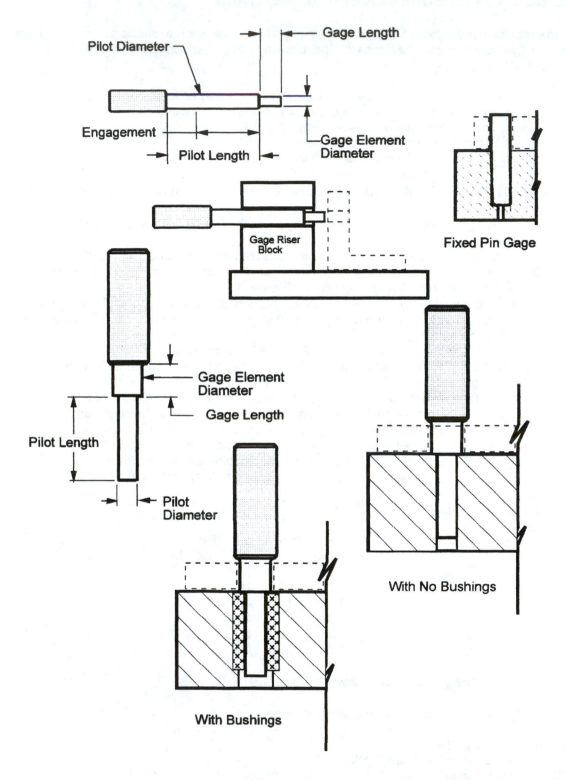

Figure 1-9 Fundamental gaging elements .

18

TOLERANCE AND FITS STUDY

Using ANSI B4.1 (U.S. customary units), let us look at a typical bore/shaft fit relationship and observe the results. From B4.1, we choose an R6 (medium running fit) for a 1.000 inch bore and shaft.

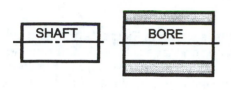

1.000 nominal diameter

Ref. ANSI B4.1

SIZE	CLASS RC6		
	TOLERANCE 0.000		
	CLEAR	BORE H9	SHAFT e8
1.000	1.6 4.8	+2.0 0	-1.6 -2.8

Expressed as Class Fit   Bore   1.000H9
Shaft   1.000e8

Unilateral Tolerance   Bore $1.000 \begin{smallmatrix} +.002 \\ -.000 \end{smallmatrix}$

Shaft $1.000 \begin{smallmatrix} -.0016 \\ -.0028 \end{smallmatrix}$

Limits   Bore $\dfrac{1.000}{1.002}$

Shaft $\dfrac{.9984}{.9972}$

In the product development/manufacturing process, all areas involved share the above tolerances. For example, gaging and inspection have 10% per B4.4; tooling and fixturing will use up about 15% to 20% as previously discussed. This distribution will therefore approximate the following values:

Shaft tolerance allowance   .0012   10% gaging   .00012
20% tooling   .00024
70% manufacturing   .00084

Bore tolerance allowance   .002   10% gaging   .0002
20% tooling   .004
70% manufacturing   .0014

The next step is to apply geometric controls of form, orientation, profile, etc. As a general guideline, these tolerances should not exceed one-half the corresponding size tolerance of a feature. For example:
With a total bore tolerance of .0012, we could apply a straightness control to the surface elements.

—	.0006

SURFACE ELEMENTS

With a total shaft tolerance of .002, we could apply a straightness control to the surface elements.

—	.001

or

⚡	.001	A

SURFACE ELEMENTS

19

The next logical consideration for the part is feature surface texture (finish). Normally, the surface texture roughness will not exceed 10% of the feature size tolerance. For example, if the total shaft size tolerance is .0012, then 10% would be .000120 mu. in. Ra max. or $\overset{120}{\sqrt{}}$. (Generally, this value is rounded down to a preferred number, such as $\overset{64}{\sqrt{}}$.)

More discussion on surface texture can be found later in the book, and further information may be found in ANSI B46.1 and ANSI Y14.36 Texture Symbols.

Application

The application of the above principles to two parts is shown in Figure 1-10. Using unilateral tolerancing helps to illustrate the fits and tolerance condition that exists at MMC. From the calculations shown, we see that the minimum clearance (both parts at MMC) of the mating datum features as well as the datum features is 0.5. Therefore there is a combined total location or orientation tolerance of 1.0 available. This tolerance may be distributed among the features of both parts in any combination that does not exceed 1.0. For this example, we have the tolerance distributed equally between both parts.

If it were possible to make perfect parts, the fit of the two parts might look like Figure 1-10b. The positional tolerance would be evenly distributed, and the axes would be aligned. It is not possible to make perfect parts, and an error of alignment can exist, illustrated by Figure 1-10c. The amount of clearance at the feature surface at MMC is equal to the total tolerance of those features. Figure 1-10c represents both parts at MMC using all available tolerance, thus illustrating a worst-case assembled fit.

Further tolerance is allowed if one or more features and/or datum features depart from the MMC size. Figure 1-11a illustrates a possible combination of LMC sizes and resulting bonus tolerances. This combination of LMC sizes results in the loosest fit possible. Figure 1-11b further illustrates a combination of tolerances, including orientation (parallelism) error, which could occur at LMC yet allow assembly of the two parts.

These principles may be applied to assemblies of three aligned diameters, as shown in Figure 1-12. For multiple-diameter assemblies, with all diameters free to float, the tolerance and resulting worst-case fit are found in the same fashion as with two-diameter assemblies. The total tolerance is distributed among all the features (both parts). If both feature diameters are offset to maximum allowed location tolerance at MMC, a line fit condition would result, allowing no additional tolerance from the datum feature. See Figure 1-12b.

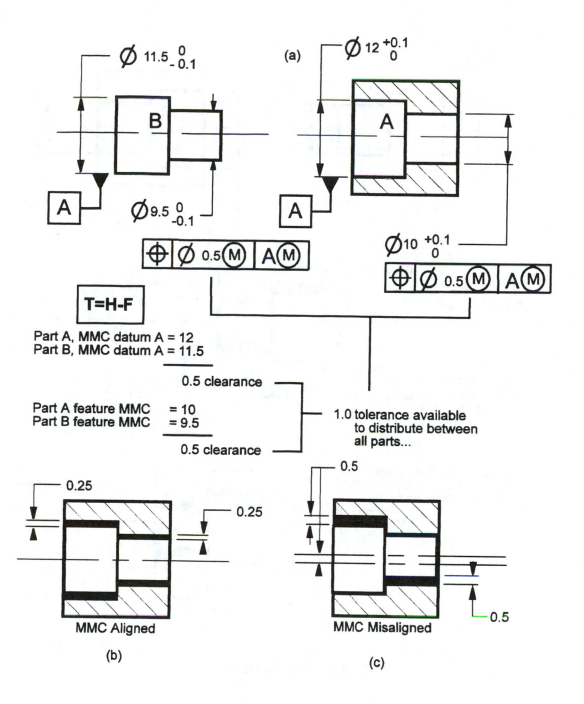

Figure 1-10 Assembly of two parts.

21

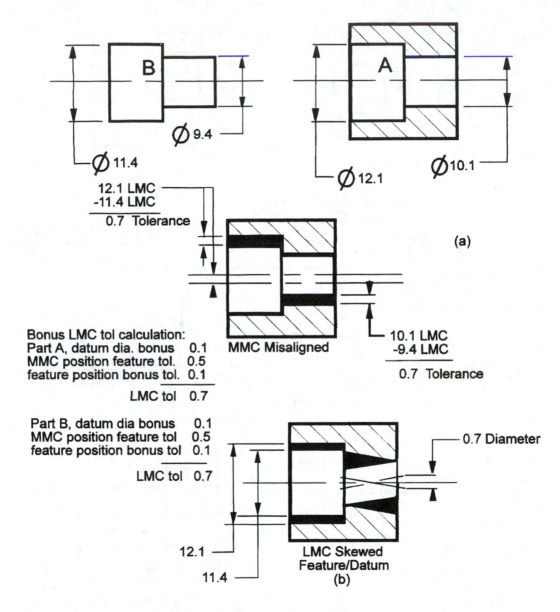

Figure 1-11 LMC fit conditions.

22

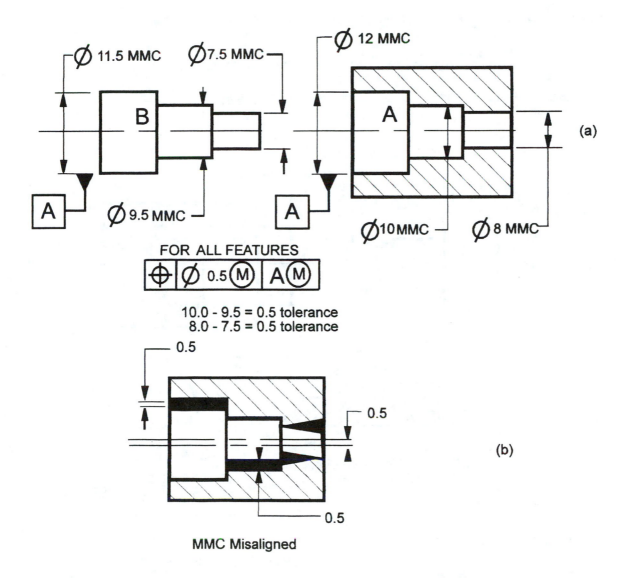

Ø 11.5 MMC   Ø7.5 MMC

B

A

Ø 9.5 MMC

Ø 12 MMC

A

(a)

Ø10 MMC   Ø 8 MMC

FOR ALL FEATURES

⊕ | Ø 0.5 Ⓜ | A Ⓜ

10.0 - 9.5 = 0.5 tolerance
8.0 - 7.5 = 0.5 tolerance

0.5

0.5

0.5

0.5

MMC Misaligned

(b)

**Figure 1-12 Assembly of three
aligned diameters.**

23

In Figure 1-13, the datum is a threaded feature that will "lock" the assembly in place. The available tolerance must be halved and distributed between all diameters of both parts. We will do more on this later in the section on Position tolerance; this exercise serves to introduce us to the tolerancing principles. A datum feature that is a feature of size (shaft or bore) may also have a location or orientation consideration relative to the total assembly. If the datum features involved create a fixed fit (press, line, or threaded fit), they must be using up some of the tolerance that will be unavailable to other features. "Datum float" will not occur.

The free assembly formula is:

tolerance = bore - shaft
        or T = H - F

The fixed assembly formula is:

tolerance = $\dfrac{\text{bore - shaft}}{2}$ or $T = \dfrac{H - F}{2}$

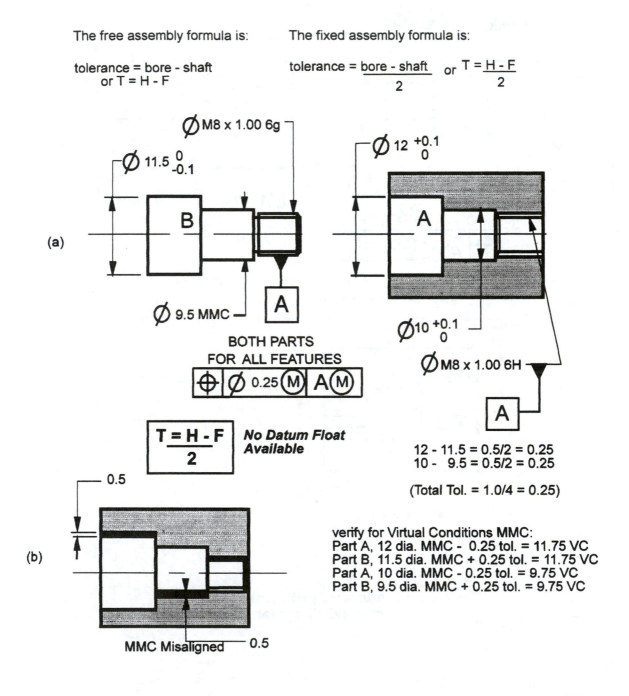

12 - 11.5 = 0.5/2 = 0.25
10 - 9.5 = 0.5/2 = 0.25

(Total Tol. = 1.0/4 = 0.25)

verify for Virtual Conditions MMC:
Part A, 12 dia. MMC - 0.25 tol. = 11.75 VC
Part B, 11.5 dia. MMC + 0.25 tol. = 11.75 VC
Part A, 10 dia. MMC - 0.25 tol. = 9.75 VC
Part B, 9.5 dia. MMC + 0.25 tol. = 9.75 VC

Figure 1-13   Threaded feature datum.

24

TOLERANCE EXPRESSION

ASME Y14.5 states that when using limit dimensioning, the high limit (maximum value) is placed above the low limit (minimum value), and when tolerance is applied in a line (.740-.760), the low limit is expressed first, regardless of the feature type (internal versus external).

This practice appears awkward, and it can be hard to remember. Further, because MMC conditions give closest fit and are the basis for functional gaging designs, it would seem to make sense to show the MMC limit as the "top" or "first" dimensional value. In addition, many tool builders and machinists work to the MMC limits first so as to the most remaining tolerance available for tool wear, salvage, or rework. For these reasons, we'll express the MMC limits *first* in most examples throughout this book where limit dimensions are used.

Placing the MMC limit above or first is not in accordance with ASME Y14.5, but does highlight the MMC values of mating features and helps us think more of mating fits and preferred numbers given in ANSI B4.1 and B4.2. This approach is intended as a training tool and is not a recommendation or for any other purpose. Tolerance conventions from ASME Y14.5 relative to the number of decimal places shall be followed.

Millimeter Tolerance

Where millimeter dimensions are used, the following applies:

Unilateral Tolerancing If either the plus or minus value is nil, a single zero is shown without a plus (or minus) sign.

$$50 \begin{smallmatrix} 0 \\ -0.02 \end{smallmatrix} \quad \text{or} \quad 50 \begin{smallmatrix} +0.02 \\ 0 \end{smallmatrix}$$

Bilateral Tolerancing Both the plus and minus values have the same number of decimal places, using zeros as necessary.

$$50 \begin{smallmatrix} +0.25 \\ -0.10 \end{smallmatrix} \quad \text{not} \quad 50 \begin{smallmatrix} +0.25 \\ -0.1 \end{smallmatrix}$$

Limit Dimensioning Zeros are added to either the maximum or minimum values for uniformity (following the decimal point).

$$\frac{50.25}{50.00} \quad \text{not} \quad \frac{50.25}{50}$$

Inch Tolerance

Where inch tolerances are used, both the minimum and maximum limits, as well as plus or minus values, are expressed in the same number of decimal places:

$$\frac{.502}{.500} \quad \text{not} \quad \frac{.75}{.748}$$

$$.500 +/- .005 \quad \text{not} \quad .50 +/-.005$$

$$.500 \begin{smallmatrix} +.005 \\ -.005 \end{smallmatrix} \quad \text{not} \quad .500 \begin{smallmatrix} +.005 \\ -0 \end{smallmatrix}$$

$$30.0^{\circ} +/- .2^{\circ} \quad \text{not} \quad 30^{\circ} +/- .2^{\circ}$$

Basic Dimensions/ Feature Control Frames
Basic dimensions and respective feature control frames follow the above rules...

inch | .500 | with | ⊕ | ⌀.005Ⓜ | A | B | C |

metric | 25 | with | ⊕ | ⌀0.25Ⓜ | A | B | C |

Dimensional limits are absolute and imply infinite zeros. Thus 12.2 means 12.20...0 15.0 means 15.00...0 15.01 means 15.010...0

GENERAL RULES

Three general rules are important to the further principles in ASME Y14.5 and should be reviewed carefully. Rule 1 does not apply to commercial stock or nonrigid parts as defined in Y14.5 section 6 and as shown in this text, Chapter 9. Note also that Rule 1 may be avoided or disclaimed as shown in Figure 1-15. Rule 1 does not control the relationship of features, as shown in Figure 1-16. ISO does not utilize Rule 1, but accomplishes the same result by the use of the symbol for "envelope" *circle E* per ISO 8015.

Rule 1

When only a size tolerance is specified, the limits of the dimension of an individual feature of size controls the form as well as the size of the feature.

    A. No element of the feature may extend beyond the envelope of perfect form at MMC.
    B. The actual local size (at any section) must be within the LMC size limit.
    C. The form control of (A) above does not apply to:
        Commercial stock: bars, sheet, tubing, etc.
        Nonrigid parts subject to variation in the *free state*.

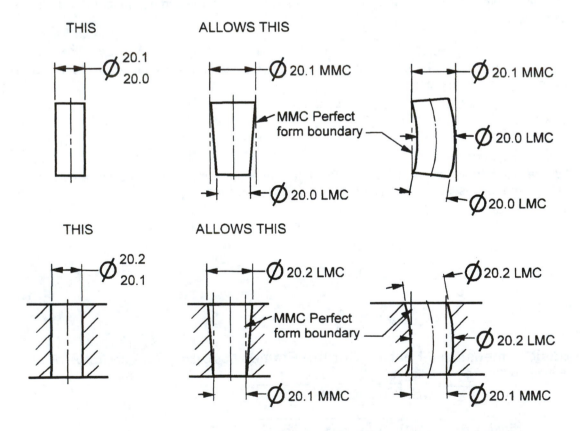

Figure 1-14 General Rule 1.

Features Subject to Size Variation:

- No element of the actual feature shall extend beyond a bound~~~ ~~ ~ct form at MMC.

- The actual size of the feature at any cr~~~ limit of size.

- EXE~~~

WHEN IT IS NECESSARY TO ALLOW A SURFACE OR SURFACES TO EXCEED THE BOUNDARY OF MMC FORM NOTED IN RULE #1, A NOTE SUCH AS " PERFECT FORM AT MMC NOT REQUIRED" IS ADDED TO THE DRAWING FOR THE FEATURE OR SURFACE NOTED.

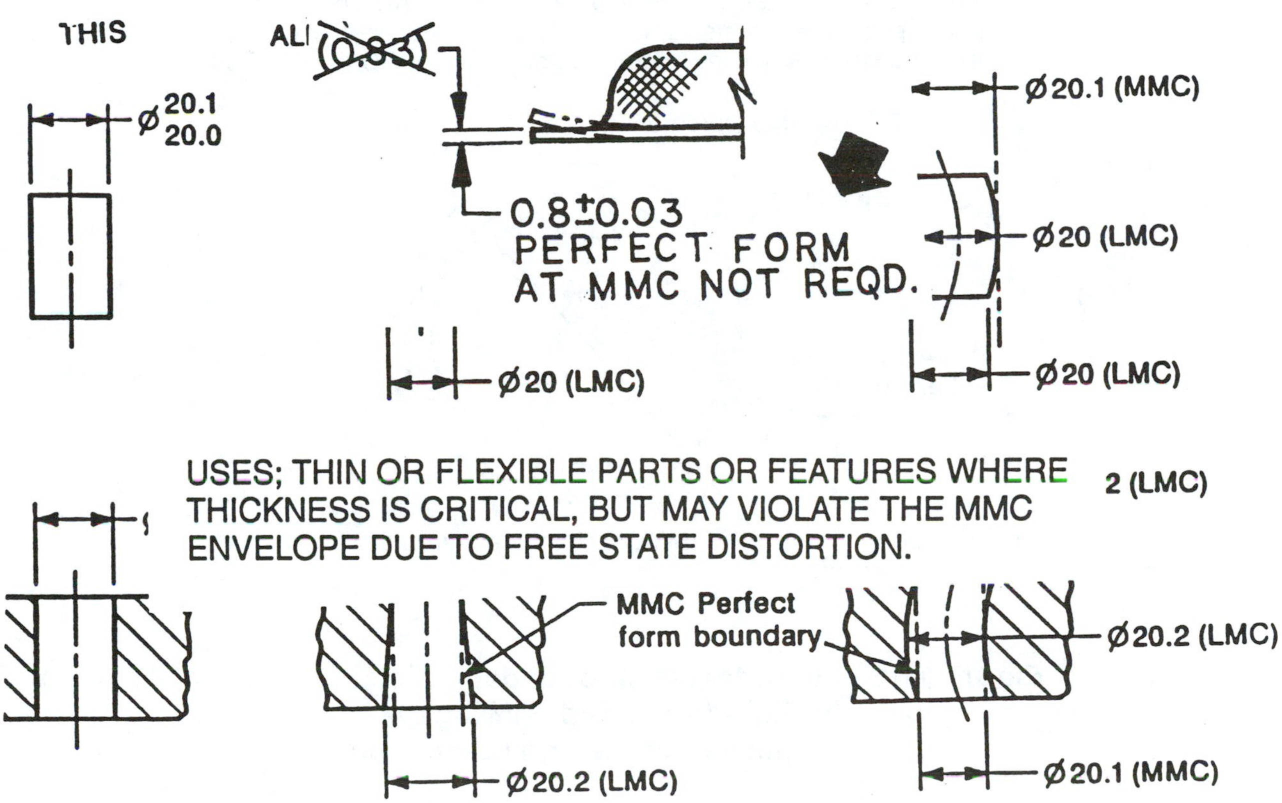

~~~ is covered by

~~~ation in the unrestrained condition.

THIS   AL~~~

$\emptyset^{20.1}_{20.0}$

0.8 ± 0.03
PERFECT FORM
AT MMC NOT REQD.

$\emptyset$ 20 (LMC)

$\emptyset$ 20.1 (MMC)

$\emptyset$ 20 (LMC)

$\emptyset$ 20 (LMC)

USES; THIN OR FLEXIBLE PARTS OR FEATURES WHERE THICKNESS IS CRITICAL, BUT MAY VIOLATE THE MMC ENVELOPE DUE TO FREE STATE DISTORTION.

2 (LMC)

MMC Perfect form boundary

$\emptyset$ 20.2 (LMC)

$\emptyset$ 20.2 (LMC)

$\emptyset$ 20.1 (MMC)

Figure 1-15 Rule 1 Exceptions.

27

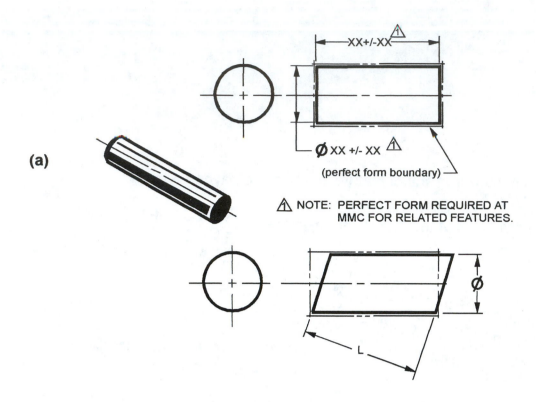

(a)

XX+/-XX

Ø XX +/- XX

(perfect form boundary)

⚠ NOTE: PERFECT FORM REQUIRED AT MMC FOR RELATED FEATURES.

Ø

L

Caution should be taken when evaluating hole diameters via two point measurements. It is possible to obtain in-spec MMC measurents, yet the MMC gage pin may not GO thru the hole.

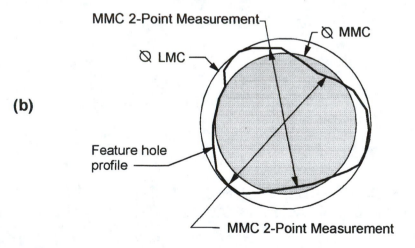

(b)

MMC 2-Point Measurement

Ø̷ MMC

Ø̷ LMC

Feature hole profile

MMC 2-Point Measurement

Figure 1-16 (a) Perfect form boundary
(b) Rule 1 explored: MMC gage
pin versus two-point measurement.

Rule 2

Rule 2 (Figure 1-17) has been revised from earlier standards. Some brief history includes:
USASI Y14.5-1966 implied that MMC applied to all tolerances of position.
ANSI Y14.5-1973 allowed for optional interpretations. (confusing)

ANSI Y14.5M-1982 required the use of a modifier symbol Ⓜ Ⓛ or Ⓢ for positional tolerances.

In the ISO system, all geometric controls are and have always been implied at Regardless of Feature Size (RFS). With the revision to Rule 2, the ASME and ISO standards are in agreement relative to this principle. The rules for screw threads and gears and splines remain unchanged.

Rule 2: For all features of size, RFS applies with respect to the individual tolerance, datum reference or both, where no modifing symbol is specified. MMC or LMC must be specified on the drawing when that is the requirement.

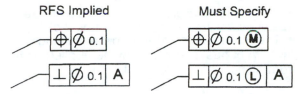

Screw Thread Rule

Each geometric tolerance control (form runout, etc)
and/or datum reference specified for a threaded feature applies to the
Pitch Diameter.. words such as MAJOR DIA or MINOR DIA must be
added beneath the feature control frame if needed.

For gears and splines, qualifying note must be added beneath the control frame, as there is no implied meaning.

Figure 1-17 Rule 2 and screw thread/gear spline rule

MATERIAL CONDITION

Even though MMC was used in the Chevrolet Drafting Handbook of the 1940s, consideration of feature material Condition (size) has generally not been widely applied in determining tolerance necessary for part acceptance.We have been accustomed to applying tolerances at face value, such as +/-.010, regardless of size. There are three material conditions to consider, MMC, LMC, and RFS. MMC and LMC have symbols, whereas RFS does not, as it is now implied, unless otherwise specified (Rule 2). When applied to feature controls, MMC and LMC will allow a bonus tolerance as the feature departs from its MMC or LMC size limits. This amount of departure may be added to the specified geometric tolerance as a bonus tolerance (more on bonus tolerance later). RFS is applied to coordinately dimensioned drawings, because the coordinate system has no provisions for applying bonus tolerances

Figure 1-18 illustrates how MMC or LMC is, or is not, applied to various form, orientation, profile, location, or runout controls. In developing a good design, it is wise to consider size controls, and the limits imposed by Rule 1 first, then consider controls of form, orientation, profile, and runout. For application of symbols, see Figure 1-19.

Regardless of Feature Size (RFS)

Symbol: none
The condition where a geometric tolerance for a feature of size applies (unaltered) at any size increment within the size tolerance.

Maximum Material Condition (MMC)

Symbol: (M)
The condition where a geometric tolerance for a feature of size applies when the feature contains the maximum amount of material. Example: The high limit for a shaft or pin and the low limit for a hole or slot.   A bonus tolerance may be applied as the feature departs from MMC.

Least Material Condition (LMC)

Symbol: (L)
The condition where a geometric tolerance for a feature of size applies when the feature contains the least amount of material within the stated size limit. LMC is the opposite of MMC. A bonus tolerance may be applied as the feature departs from LMC.

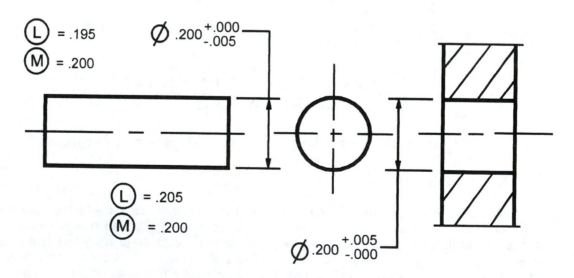

Figure 1-18 Material condition.

CHARACTERISTIC	MATERIAL CONDITION GUIDE APPLICATION of Ⓜ or Ⓛ
STRAIGHTNESS ⎯	NA, if a feature surface, must be within size limits. YES, if a feature axis or centerplane.
FLATNESS ⟋⟍	NA, when associated with a size dim. the ⟋⟍ must be less than size tol.
CIRCULARITY O	NA. Tolerance zone is at the surface. O must be within size limits. Exception: if AVG DIA & Ⓕ applied.
CYLINDRICITY ⌭	NA. Same as Circularity.
PROFILE LINE/ ⌒ SURFACE ⌓	NA. Tolerance zone is at the surface and must be within size limit when applicable.
ORIENTATION ⊥ ∥ ∠	NA if a feature surface, must be within size limits. YES if for a feature axis or centerplane.
POSITION ⊕	YES
CONCENTRICITY ◎	NO
SYMMETRY ≡	NO
RUNOUT ↗ ⤭	NO

Figure 1-19 Material condition application.

INTRODUCTORY CONCEPTS SUMMARY

Symbols have a worldwide meaning and application.

Symbols can be combined into control frame "graphic sentences."

Tolerance values should be logical, justifiable, and defendable.

"Fundamental Rules of Dimensioning" and "Definitions" are the foundations of ASME Y14.5.

Preferred limits and fits of ANSI B4.1 and B4.2 should be considered where possible.

Rule 1 does not apply to feature relationships.

For features of size, consider limits imposed by ASME Y14.5M Rule 1 before form/orientation controls are added.

All drawing tolerance limits are not available to the manufacturing processes.

RFS is implied for all feature controls. MMC or LMC must be specified where needed, per Y14.5M Rule 2.

ISO 8015 Ⓔ
Drawings from outside the United States may be generated per ISO standards, which normally employ the "Principle of Independency" of ISO 8015. To accomplish the same result of Rule 1, ISO uses the circle E symbol.

MMC and LMC can impact tolerance application and can only be applied to features of size with an axis or centerplane.

Major industrial country dimensioning and tolerancing standards are as follows:

ISO	1101, 2692, 5458, 7083, 8015
AUSTRALIA	AS 1100.201
CANADA	CSA B78.2
FRANCE	NFE04-121
GERMANY	DIN 7184
ITALY	UNI 7226
JAPAN	JIS B0021
UNITED KINGDOM	BS 308(3)
UNITED STATES	ASME Y14.5M-1994

EXERCISE 1 General Dimensioning (Introductory Concepts)

1. Rule 1 controls: a. feature local size and size envelope
                     b. the relationship of features
                     c. local form tolerance of a feature

True or False:
2. Rule___ implies RFS is applied to Profile and Runout controls. T F

3. Prior to ANSI Y14.5- 1973, MMC was implied for position controls. T F

4. ISO drawings imply MMC for Position controls. T F

5. All dimensions on a drawing must have a tolerance. T F

6. Drawing callouts referring to screw threads apply to the pitch diameter. T F

7. For features of size, the limits of size and Rule 1 are generally applied before form or orientation controls are added. T F

8. ISO 8015 Independency Principle is the same as ASME Y14.5 Rule 1. T F

9. Rule 2 states that RFS is implied for tolerance controls. T F

10. Identify the figures in the feature control frame:

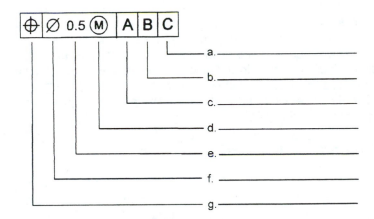

```
a. _____
b. _____
c. _____
d. _____
e. _____
f. _____
g. _____
```

11. Identify each of the geometric characteristic symbols below:

33

2 FORM TOLERANCE CONTROLS

Straightness	Flatness	Circularity	Cylindricity
—	▱	○	⌭

Individual Feature Controls, No Datum Relationship

Now that we have a general understanding of fundamental dimensioning and tolerancing principles, we will start with the most basic of the geometric controls.

The most elementary element is a *point* in space. The connection of two points will create a *line*, whereas a line and a third point (or three points) will create a *plane*. The intersection of two planes forms a *line (centerline or axis)*. Points equidistant from an axis create a *circle*, and two circles along the same line create *a cylinder*. These descriptions constitute the most elementary form controls of the ASME Y14.5 standard, and are classified as *Tolerances of Form* .

The common link between all form controls is that they are controls for individual features with no further relationships or datum reference. These controls, as noted, are easy to define but difficult to measure or evaluate, because they have no origin of reference (or datum). Form controls are considered a refinement of size, and when applied to a feature surface element, they must be contained within the limits of size (per Rule 1) unless stated otherwise.

STRAIGHTNESS

Straightness is a condition where *an element of a surface or axis is a straight line*. A straightness control specifies a tolerance zone within which the axis (derived median line) or surface element of the feature must lie. The control of *surface elements is two dimensional* (height and length), whereas the control of a cylindrical feature axis yields a *three-dimensional cylindrical* tolerance zone. Though simple in definition, straightness is the most complex form control and the most difficult control to evaluate or verify. It is important to note that straightness controls apply only in the view specified on the drawing. The placement of the feature control frame is extremely important. See Figure 2-1.

Straightness of Cylindrical Features

In Figure 2-2, the feature is a cylindrical shape. There are two possible elements to control, surface elements and the axis. In Figure 2-2a, the control frame is separated from the size dimension, and the leader line points directly to the feature surface. In Figure 2-2b, the control frame is located below and connected to the size dimension. Figures 2-2a and 2-2b represent two distinct designs. The tolerance zone for Figure 2-2a is the *space between two parallel lines at the feature surface*, whereas the tolerance zone for Figure 2-2b is a *cylindrical zone at the feature axis*. Because the tolerance zone of Figure 2-2a is a two-dimensional zone at the surface, the straightness tolerance must be contained within the size limits (per Rule 1). Because the tolerance zone of Figure 2-2b is at the feature axis (a cylindrical 3D zone), the axis may be bent within the straightness tolerance limit of .010. The combination of size controls and axial straightness controls may allow the feature to violate the boundary of perfect form at MMC (Rule 1) and therefore Rule 1 does not apply.

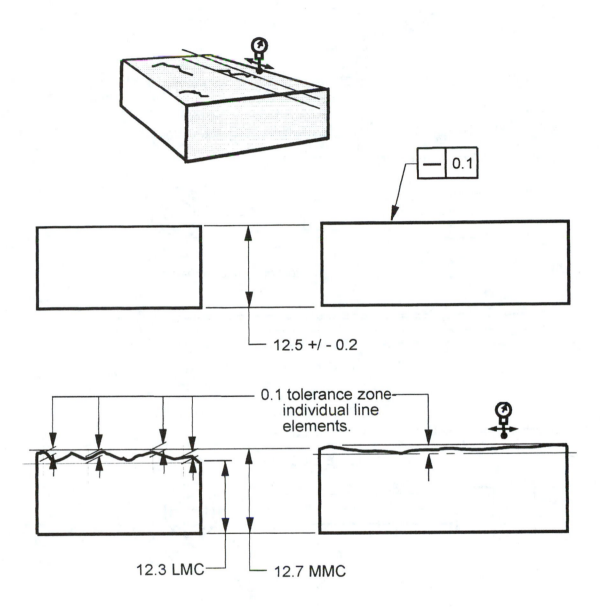

— | 0.1

12.5 +/ - 0.2

0.1 tolerance zone—
individual line
elements.

12.3 LMC   12.7 MMC

Figure 2-1 Straightness - surface line elements.

Straightness- Surface elements

$\boxed{— | .001}$

$\varnothing$ .500 +/- .003

.001 tol zone - line elements

.497 LMC local size

.503 MMC envelope

Note: Rule #1 applies

Straightness- Axis RFS

~~Rule #1~~

$\varnothing$ .500 +/- .003

$\boxed{— | \varnothing .010}$

$\varnothing$ .497 - .503 size limits

$\varnothing$ .010 axis tol zone RFS

$\varnothing$ .513 outer boundary

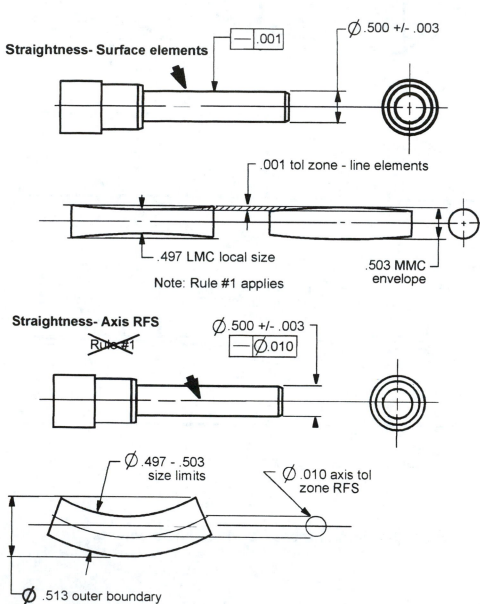

Figure 2-2   Straightness - cylindrical surface elements and axis (both RFS).

Straightness is applied RFS unless otherwise stated. With cylindrical features, the placement of the control frame can drastically change the design intent. Where straightness controls are used in *conjunction with orientation or other controls*, the straightness tolerance *shall not* be greater than the specified orientation (or other control) values. When *not used in conjunction* with other controls and when applied to a *feature axis*, the straightness tolerance may be greater than the MMC size envelope. In this instance, the MMC limits may be violated, nullifying Rule 1.

As shown in Figure 2-3, straightness may be applied to a feature (MMC) axis. In this example, the feature axis must be straight within a cylindrical tolerance zone of .010 when the feature is at the MMC size of .503. The figure illustrates that as the size departs from MMC size, a bonus tolerance can be added to the original value. The virtual mating size (collective effects of size and geometric tolerance) remains the same (.513). The bore or hole in the mating part would have to be at least .513 diameter to ensure a clearance fit.

Look again at Figure 2-2. The straightness control is applied RFS; therefore, the tolerance is a constant-diameter zone of .010. The *outer boundary,* or mating size, is variable from .513 to .507 diameter. With RFS, the tolerance is constant, whereas the mating size is variable, but with MMC, the tolerance zone is variable and the MMC mating size (virtual condition) is constant.

Straightness per Unit Length

If the design cannot tolerate an abrupt step, as shown in Figure 2-3, a tolerance per unit may be added to the control frame callout. The straightness control symbol is shown once and is applicable to both the total tolerance (top) as well as the incremental per unit straightness (bottom). This consideration may be appropriate for plungers or shafts and mating bore fits where very close tolerances or sliding fits are required.

Straightness Applied LMC

For some designs, it may be important to ensure a minimum cross-sectional area for strength reasons, such as a shaft or column under a compression load. Figure 2-4 illustrates such a case. Further, in addition to control of the axial straightness at minimum size, it may be necessary to relate this control to a datum surface. In such a case, perpendicularity or position may be the desired control. Note the resulting MMC envelope and tolerance zone that could occur.

Straightness of Centerplanes

In the past, straightness controls were applied to centerplanes, as shown in Figure 2-5. As the definition of straightness deals with line elements or axes, this application of centerplane control could be argued as incorrect. If the design intent is to control the centerplane of a feature, such as a keyslot, the use of symmetry or position may be considered. If the intent is to control surface elements, however, straightness control applied to surface elements along with symmetry or position control of the centerplane relative to a datum feature may be the proper control. The design should dictate the proper controls needed.

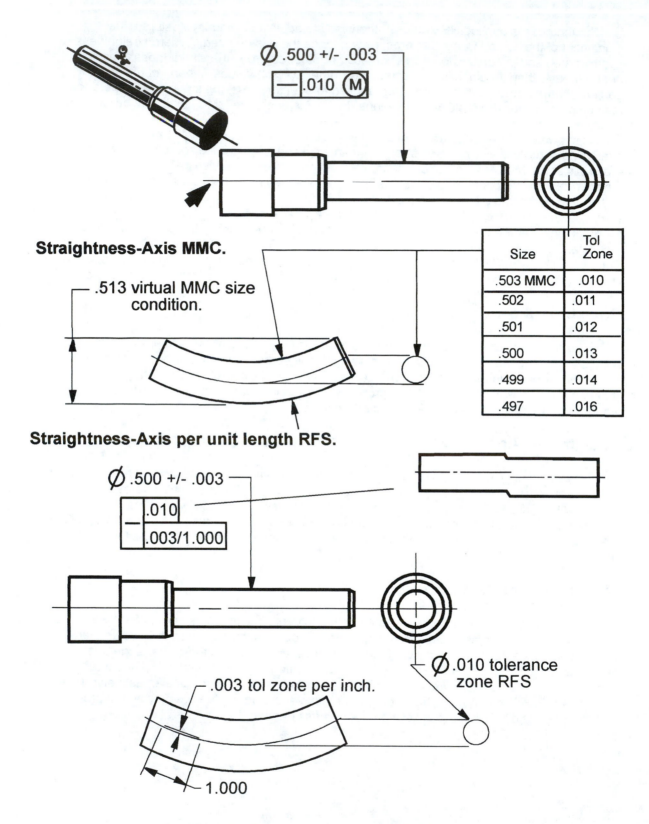

Ø .500 +/- .003

— .010 (M)

Straightness-Axis MMC.

.513 virtual MMC size condition.

Size	Tol Zone
.503 MMC	.010
.502	.011
.501	.012
.500	.013
.499	.014
.497	.016

Straightness-Axis per unit length RFS.

Ø .500 +/- .003

.010
.003/1.000

Ø.010 tolerance zone RFS

.003 tol zone per inch.

1.000

Figure 2-3   Straightness - axis MMC and per unit length.

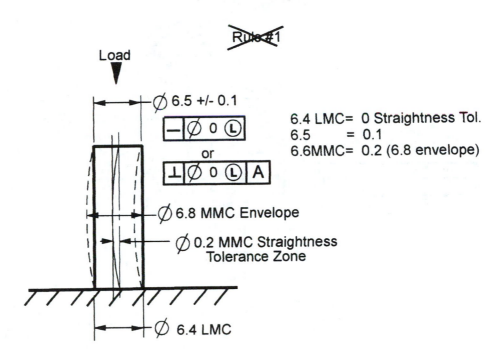

Figure 2-4 contents:

Rule #1 (crossed out)

Load

⌀ 6.5 +/- 0.1

⎓ | ⌀ 0 Ⓛ

or

⊥ | ⌀ 0 Ⓛ | A

6.4 LMC= 0 Straightness Tol.
6.5 = 0.1
6.6 MMC= 0.2 (6.8 envelope)

⌀ 6.8 MMC Envelope

⌀ 0.2 MMC Straightness Tolerance Zone

⌀ 6.4 LMC

Note: Where a straightness tolerance is used in conjunction with an orientation or position tolerance, the straightness tolerance value shall be no greater than the specified orientation (or position) tolerance value.

Figure 2-4   Straightness applied LMC.

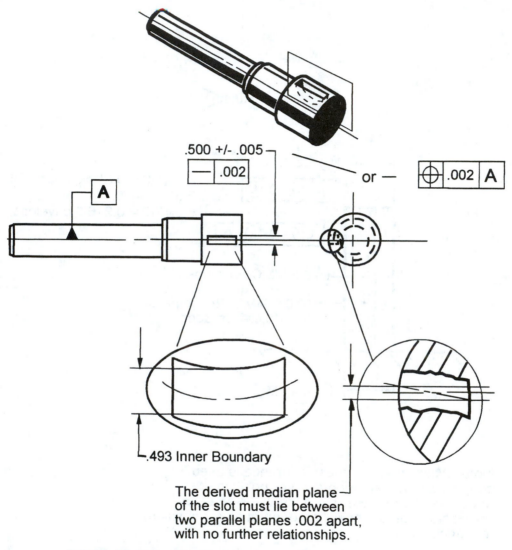

.500 +/- .005

| — | .002 |

or —

| ⊕ | .002 | A |

A

.493 Inner Boundary

The derived median plane of the slot must lie between two parallel planes .002 apart, with no further relationships.

NOTE: Position tolerancing could be used which would define the centerplane tolerance zone, as well as control alignment with datum axis A.

Figure 2-5 Straightness of centerplane.

Virtual Condition

Virtual condition is the boundary generated by the collective effects of MMC or LMC size and any geometric tolerance. With this definition in mind, the effects of virtual condition can be seen when feature controls such as straightness perpendicularity or position are applied to a feature axis or centerplane. In Figure 2-6, the straightness control as applied to the axis (.015) is additive to the size tolerance, with the total mating size virtual condition being .517 diameter. If the straightness control were eliminated, the mating size could not exceed .502, per Rule 1. The conclusion reached for Figure 2-6 is that the diameter tolerance is critical (fit), but the straightness (form) is not. This set of circumstances may exist for certain shafts of great length that may tend to bow or droop of their own weight if observed in a horizontal setup. Even though it should be understood that when a geometric tolerance control is applied to a feature axis or centerplane Rule 1 is not invoked, in this case, the addition of a note "PERFECT FORM AT MMC NOT REQUIRED" beneath the control frame will help ensure the design intent. Potential *Virtual Conditions* include:

HOLES	SHAFTS
MMC size - Geometric tolerance Virtual Condition	MMC size +Geometric tolerance Virtual Condition
LMC size +Geometric tolerance Virtual Condition	LMC size - Geometric tolerance Virtual Condition

Also ref figure 9-14.

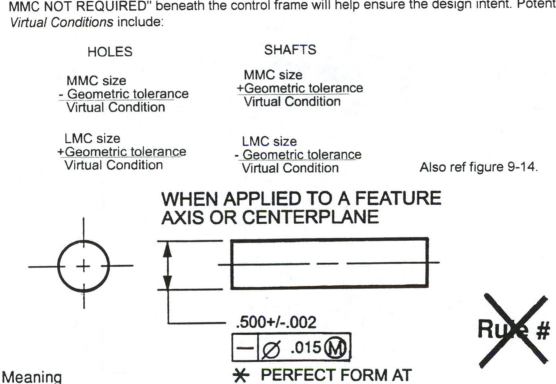

WHEN APPLIED TO A FEATURE AXIS OR CENTERPLANE

.500+/-.002

| — | ⌀ .015 Ⓜ |

Meaning

✱ PERFECT FORM AT MMC NOT REQUIRED

Rule # 1

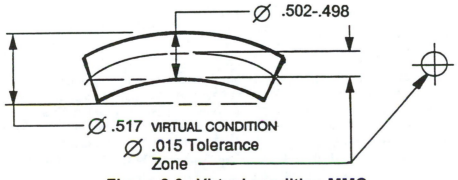

⌀ .502-.498

⌀ .517 VIRTUAL CONDITION
⌀ .015 Tolerance Zone

Figure 2-6   Virtual condition MMC

41

FLATNESS

Flatness differs from straightness in that flatness controls *all surface elements* (multi-directional) of a planar surface; see Figure 2-7. A Flatness tolerance control specifies a tolerance zone confined by two parallel planes within which the *entire surface* must lie. Compared with straightness, flatness is relatively straightforward, since it can only be applied to a single planar surface. The location of the control frame is not critical, and there can be only one interpretation; thus the control frame may be attached as a flag to the feature extension line or may be directed to the feature surface by a leader line. Once defined and verified, these planar surfaces often become *primary datum surfaces*.

If the design will not tolerate abrupt changes or steps within the flatness tolerance, it may be necessary to apply a *flatness per unit area*. Any area or zone may be used for this incremental verification, with the total flatness applied to the entire surface. The location of the feature control frame is optional and has only one interpretation, as above.

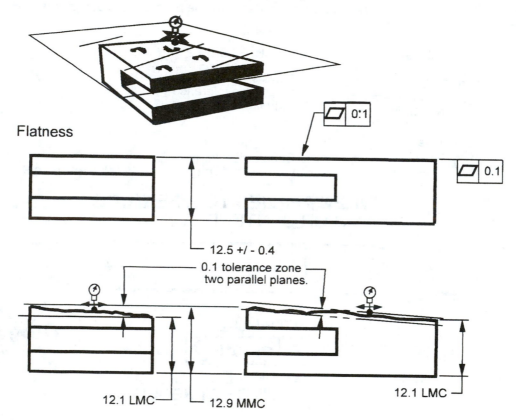

Figure 2-7 Flatness

Waviness

Waviness describes secondary surface texture. Waviness is often confused with flatness, but it may be described as the height and length of surface undulations that occur due to the processing stresses, pressures and temperatures. Waviness is described in ANSI B46.1 Surface Texture and is contained within the flatness tolerance. Waviness may be related to the swells at sea, which rise and fall but are contained within the high/low tides. See Figure 2-8. Refer to ANSI B46.1 and ANSI Y14.36 for more on surface texture and the symbols involved.

Flatness per Unit

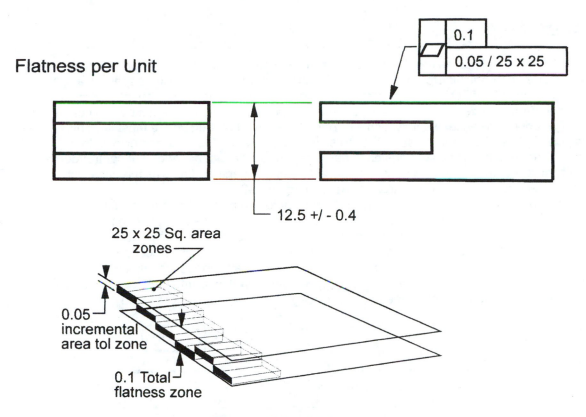

0.1
0.05 / 25 x 25

12.5 +/ - 0.4

25 x 25 Sq. area zones

0.05 incremental area tol zone

0.1 Total flatness zone

Figure 2-7a Flatness per unit area

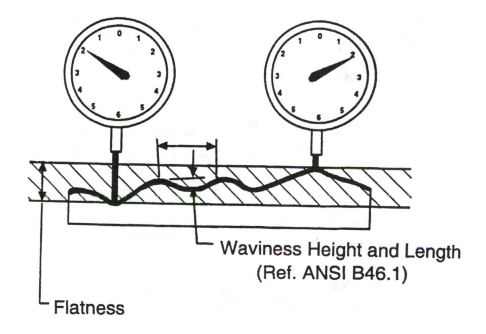

Waviness Height and Length
(Ref. ANSI B46.1)

Flatness

Figure 2-8 Waviness .

CIRCULARITY

Circularity is defined as the condition of a surface where all points of the surface intersected by any plane perpendicular to an axis are equidistant from that axis, or, with respect to a sphere, where all points of the surface intersected by any plane passing through a common center are equidistant from that center. A circularity tolerance zone is the space bounded by two concentric circles (a radial difference) within which each circular element of the surface must lie. This definition applies RFS, regardless of feature size. Circularity is a form control for circular or spherical features and is two dimensional, because the tolerance zone does not control depth. The placement of the control frame is not critical to design intent, and as was the case with flatness, there can be only one interpretation. See Figure 2-9.

Note that the circularity error may not be readily detected by caliper or micrometer type measurement on shafts with an odd number of lobes. This type measurement can reflect a constant diameter but cannot detect a radial error. See Figure 2-10. It is important to recognize that vee blocks are not always reliable in circularity measurement, because certain angles will hide the lobing. Lobing is most commonly found in ground shafts; lobes may number between 3 and 19. Further, if more than one source or supplier is involved, care must be taken to ensure that all involved are using similar methods, or quality levels may be compromised.

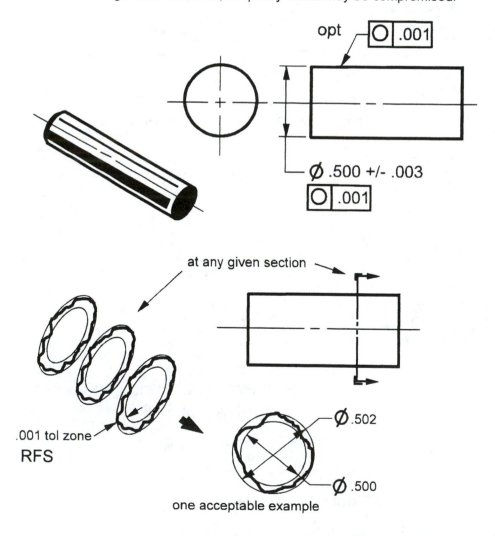

Figure 2-9 Circularity (RFS).

44

(a) When applied to a sphere, a circularity
   control specifies a tolerance zone that
   is the space between two concentric circles
   lying in any plane that passes thru the center
   of the sphere.

(b) When applied to a cylinder, torus, or cone,
   a circularity control specifies a tolerance
   zone that is the space between two concentric
   circles lying in any plane normal (perpendicular)
   to the feature spine at any selected location.

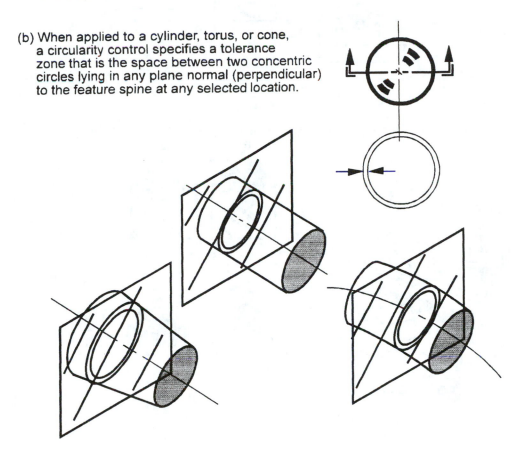

Figure 2-9 (continued)

Review Figure 2-11. Without a circularity control specified, the conditions shown, allowed by Rule 1, result in a MMC envelope of .860 diameter and a cross section of .840 diameter. The diameters may be displaced, which could result in a D-shaped shaft portion, as shown. To remove this possibility, we should consider the use of the circularity control as shown in Figure 2-12. It is important to note that the circularity control elements at the various locations do not have to line up.

Circularity is the only *form control* with an additional standard concerning assesment methods. ANSI B89.3.1 covers circularity measurement methods for critical features, not only when the design limits are required on the drawing, but when the precise methods of measurement become integral to the design specifications. Not only are feature tolerances given; the measurement methods are also shown in the control frame, including technique, filter response, and stylus tip radius. See Chapter 9 of this book and ANSI B89.3.1 for more information.

CALIPER AND MICROMETER WILL
NOT PICK UP ROUNDNESS ERROR

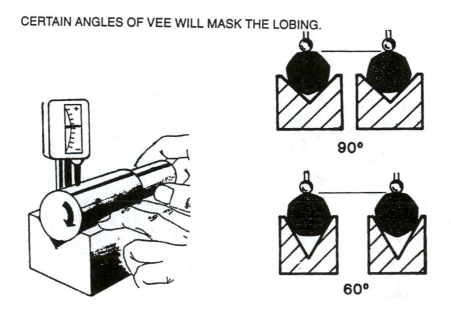

CERTAIN ANGLES OF VEE WILL MASK THE LOBING.

90°

60°

**Figure 2-10 Caliper, micrometer, and
vee locator measurement.**

CYLINDRICITY

Cylindricity is a condition of a surface of revolution in which *all points of the surface* are equidistant from a common axis Cylindricity control specifies a tolerance zone bounded by two concentric cylinders, within which all elements of the surface must lie. Cylindricity is the control that pulls *circularity, straightness, and taper* together to form a *concentric cylinders* tolerance zone. This control is considered somewhat difficult to measure with conventional equipment because the tolerance zone is three dimensional, similar to flatness, and one can argue that *all elements* cannot be measured economically. But the B89 quality series standards gives the quality engineers some credit for *exercising judgment* in this regard. Generally, enough points are to be measured so as to give *reasonable assurance* the specification has been met. This reasoning is a given with all evaluation systems.

The placement of the control frame is optional, as with circularity. See Figure 2-13.

46

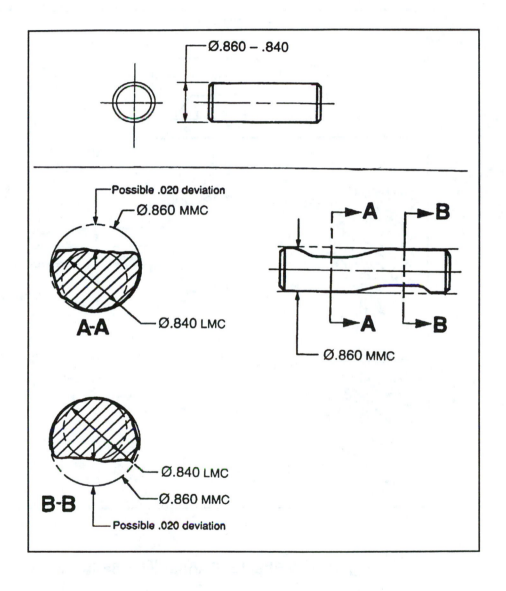

Figure 2-11 Effects of Rule 1 on circularity.

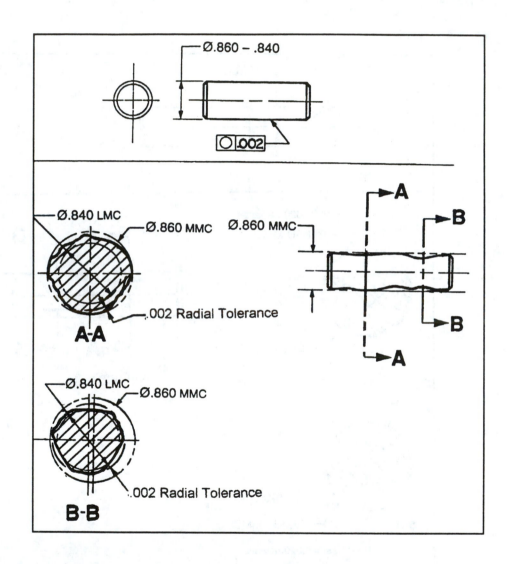

Figure 2-12 Effects of added circularity control.

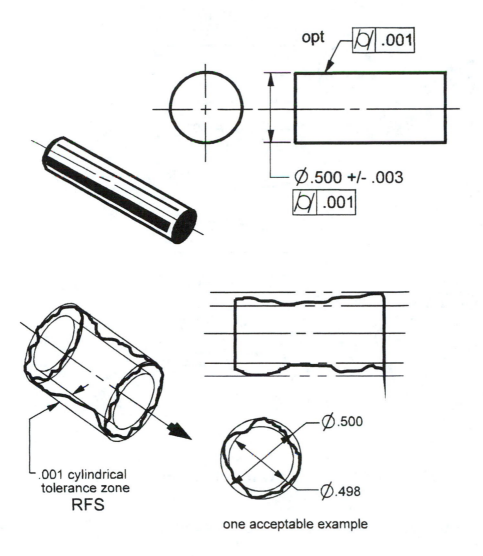

opt $\boxed{\cancel{\bigcirc} \ | \ .001}$

$\varnothing$.500 +/- .003

$\boxed{\cancel{\bigcirc} \ | \ .001}$

.001 cylindrical
tolerance zone
RFS

$\varnothing$.500

$\varnothing$.498

one acceptable example

Figure 2-13   Cylindricity.

Computerized equipment may be programmed to calculate cylindricity based on probe-type measurement using a formula derived from the square-root sum of the measurements, such as straightness squared, circularity squared, and taper squared. This technique is illustrated in Figure 2-14. In theory, if measured values are all .001 as shown, a program may calculate a cylindricity value of .0017. If the program only uses circularity and straightness in the calculation, a value of .0014 may evolve. Further, the rounding practices used may impact the acceptance or rejection decisions on parts. Consistency is the necessity, and investigating equipment programs may be in order.

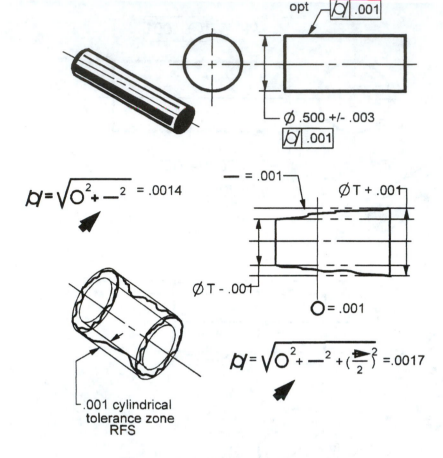

opt $\boxed{\cancel{O}\ \ .001}$

$\emptyset$ .500 +/- .003

$\boxed{\cancel{O}\ \ .001}$

$$\cancel{O} = \sqrt{O^2 + {—}^2} = .0014$$

— = .001

$\emptyset$ T + .001

$\emptyset$ T - .001

$O = .001$

$$\cancel{O} = \sqrt{O^2 + {—}^2 + \left(\frac{\blacktriangleright}{2}\right)^2} = .0017$$

.001 cylindrical
tolerance zone
RFS

Figure 2-14 Cylindricity measurement.

FORM TOLERANCE CONTROLS SUMMARY

Surface elements are controlled by circularity and straightness, whereas surfaces are controlled by cylindricity and flatness. Circularity and straightness are two-dimensional controls, whereas cylindricity and flatness are three-dimensional controls.

Circularity, cylindricity, and flatness controls apply only at feature surfaces; therefore, they cannot be modified MMC or LMC.

Straightness controls may be applied to feature surface line elements or to feature axes or centerplanes.

When circularity or cylindricity controls are used, they are to be no more than 1/2 size tolerance values.

When used in combination with orientation or position controls, straightness control shall be within the specified orientation or position control.

When specifying straightness of a feature axis, be aware that the effects of *virtual condition* may allow violation of the MMC envelope boundary.

50

When violation of the MMC envelope is the desired intent, such as features with critical thickness but noncritical straightness or flatness, the following note should be added. PERFECT FORM AT MMC (LMC) NOT REQUIRED.

From ISO 1101. (1) The direction of the ideal line or surface or position of the two concentric circles or coaxial cylinders determined by means of regression features. (2) The line or surface may be determined by the sum of the squares of distances d_1 of the measured points from a true line or circle (cylinder), as shown below.

$$\Sigma\, d_1{}^2 = \text{minimum separation (straightness,}$$
$$\text{circularity, cylindricity)}$$

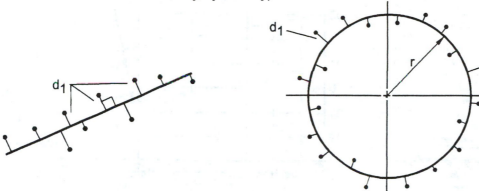

Also refer to ASME Y14.5.1M-1994 for mathematical definitions of form tolerances.

As noted previously, form controls are more easily defined than actually measured. Form controls have no datum relationships from which to set up the part. Figures 2-15 through 2-17 illustrate a few measurement options, when computerized equipment is not available or practical.

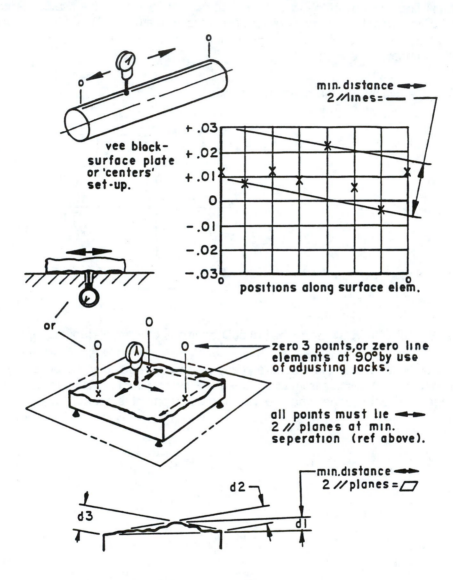

Figure 2-15 Straightness/flatness measurement practices.

52

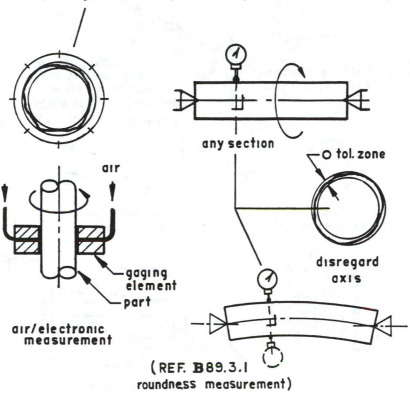

Evaluation methods include; mechanical, optical, electronic and air gaging. Roundness measuring machines which record the feature contour on polar graphs are quite widely used.

any section

○ tol. zone

disregard axis

air

gaging element

part

air/electronic measurement

(REF. B89.3.1 roundness measurement)

Figure 2-16 Circularity (roundness) measurement practices.

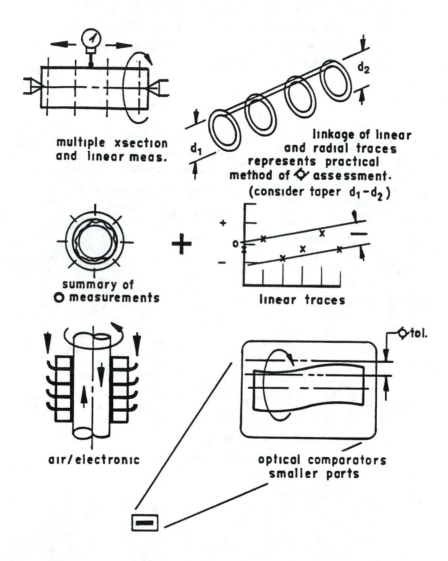

multiple xsection
and linear meas.

d_1 d_2

linkage of linear
and radial traces
represents practical
method of ◇ assessment.
(consider taper $d_1 - d_2$)

summary of
O measurements

+

linear traces

air/electronic

optical comparators
smaller parts

◇tol.

Figure 2-17 Cylindricity measurement practices.

EXERCISE 2-1 Circularity

True or False

1. Circularity is implied MMC unless otherwise specified.   T   F

2. Circularity controls apply at the _____ of a cylindrical shape.

3. Circularity controls may be applied to a tapered shaft.   T   F

4. Circularity is datum related.   T   F

5. Circularity is a refinement of size tolerance.   T   F

6. Vee blocks are best for circularity measurement.   T   F

7. Circularity tolerance zone must be within size limits.   T   F

8. The use of MMC will allow a bonus circularity tolerance.   T   F

9. Indicate a circularity control callout with a 0.1 mm tolerance
   on the figure below.   Draw the tolerance zone.

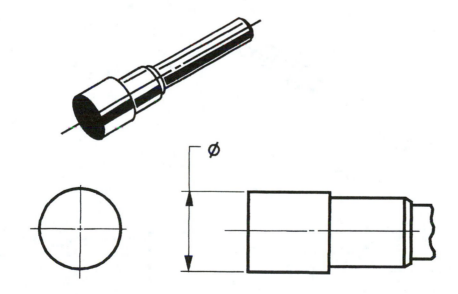

EXERCISE 2-2   Cylindricity

1. Cylindricity is implied MMC unless otherwise specified.   T   F

2. Cylindricity is applied to cylindrical shapes and is additive to size tolerances.   T   F

3. Cylindricity controls _____ and _____ of surface elements   T   F

4. Cylindricity is always datum related.   T   F

5. Cylindricity tolerance zone is the space between two concentric circles.   T   F

6. A cylindricity tolerance zone may extend beyond the MMC size envelope.   T   F

7. Cylindricity may be used to control taper, such as in the figure below.   T   F

8. Indicate a cylindricity control callout with a 0.2 mm tolerance
    on the figure below.   Draw the tolerance zone.

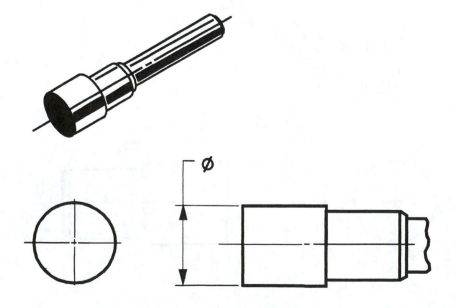

EXERCISE 2-3 Flatness

1. Flatness is applied MMC unless otherwise specified.   T   F

2. Flatness controls _____ of surface elements.

3. Flatness is additive to size tolerances.   T   F

4. Flatness is always datum related.   T   F

5. If flatness controls the entire surface, it also controls the squareness
   to a related surface.   T   F

6. The figure below may have a flatness control tolerance of 0.5.   T   F

7. Can MMC be applied to the subject surface in the figure below? Yes No

8. Indicate a flatness control callout (top surface) with a 0.2 mm tolerance
   on the figure below.   Draw the tolerance zone.

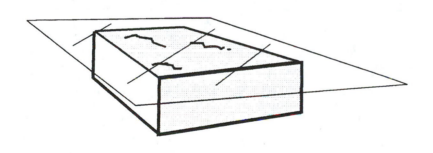

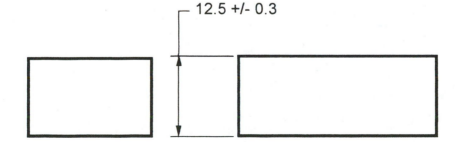

12.5 +/- 0.3

EXERCISE 2-4 Straightness

1. Straightness is implied RFS unless otherwise specified.   T   F

2. Straightness may be applied to flat surface _____, cylindrical surface _____, or_____.

3. Virtual condition is invoked when straightness is applied to a feature _____.

4. Complete the table below for the measured sizes given. Draw the tolerance zone.

Tolerance zone available if:

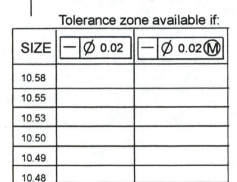

SIZE	$-$ ⌀ 0.02	$-$ ⌀ 0.02 Ⓜ
10.58		
10.55		
10.53		
10.50		
10.49		
10.48		

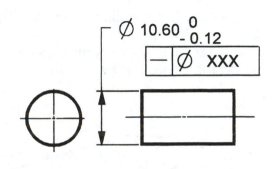

⌀ 10.60 $^{0}_{-0.12}$

| $-$ | ⌀ | XXX |

What is the *virtual mating size*_____ condition?

Complete the table below for the measured sizes given. Draw the tolerance zone.

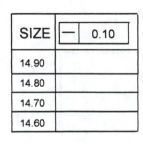

SIZE	$-$ 0.10
14.90	
14.80	
14.70	
14.60	

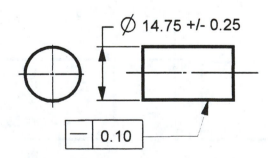

⌀ 14.75 +/- 0.25

| $-$ | 0.10 |

What is the *virtual mating size*_____ condition?

EXERCISE 2-5   Straightness

WITH STRAIGHTNESS OF SURFACE ELEMENTS: TRUE FALSE

1. Ⓜ or Ⓛ may not be applied.

2. Perfect form at MMC is required.

3. The straightness tolerance must lie within size tolerance.

4. *Virtual condition* will allow Rule 1 to be violated.

5. RFS is understood to apply.

WITH STRAIGHTNESS OF AN AXIS: TRUE FALSE

1. Datums are to be applied.

2. Rule 1 does not apply when straightness is applied to
   a feature axis or centerplane.

3. Virtual condition is equal to MMC size plus the axial
   straightness tolerance.

4. With axial straightness, the diameter symbol is optional.

5. The virtual condition mating size of a shaft that is 15 mm
   diameter +/- 0.3 mm,with an axial straightness control of 0.3
   diameter is 15.3 mm.

3 DATUMS

A *datum* is a theoretically exact point, line (axis), or plane derived from the true geometric counterpart of a specified datum feature. A datum is the origin from which the location, or other geometric characteristic, of a part feature is established. The logical progression from the control of feature surfaces, via form controls, is the use of these controlled features (points, lines, planes) as *datum features*. First, it is necessary to understand and develop a datum framework. Basic geometry reveals that three points will generate a plane, and from that plane a second plane may be constructed, perpendicular to the first, by the use of two points. A third plane, perpendicular to the other two, may be established by one point, resulting in three *mutually perpendicular planes or datum framework*. This three-plane framework is the foundation for defining three-dimensional objects. See Figure 3-1.

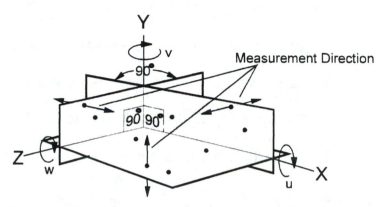

Datum Reference Framework. Three Mutually Perpendicular Planes.

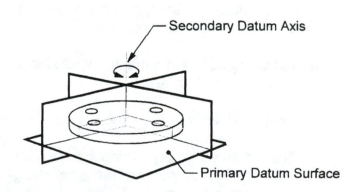

Example: Primary Surface, Secondary Axis.

Figure 3-1 Datum reference frame.

60

DATUM REFERENCE FRAME

A completely defined datum reference frame will constrain or restrict a feature or part in three translational directions (x, y, and z), and three rotational orientations (u, v, and w), where u is in rotation about the X axis, v is in rotation about the Y axis, and w is in rotation about the Z axis.

A datum reference frame need not contain controls for all degrees of freedom, for location, and for orientation of tolerance zones. For example, a datum reference frame defined by only one plane is sufficient to control perpendicularity or parallelism to *that* plane.

CYLINDRICAL DATUM FEATURES

Two sets of conditions may exist for cylindrical objects or features. The first set is a cylidrical feature of short length, where the *primary design influence* rests with the mounting surface and the *secondary influence (or tertiary influence)* is the intersecting planes that form the axis for the datum frame. See Figure 3-1. The second set of conditions is where the axis of the datum feature (of greater length), with possibly rotary or linear motion involved, exerts the *primary design influence* and is thus the primary datum. The secondary datum, which may be a point or stop, will arrest lateral float, and the tertiary datum point will stop of rotation about the axis. See Figure 3-2. In that figure, three equally spaced tool elements, or jaws, may close to isolate the feature axis and establish it as the primary design influence, while the end stops lateral movement and the keyslot stops rotational movement, locking the part in the reference framework. Generally, two issues help determine when the axis might be considered the *primary influence*:

1. The ratio of surface area to diameter length, if a static application
2. Rotary or linear movement of the cylindrical feature relative to a mating component or dynamic application

In most designs, this influence should be clear.

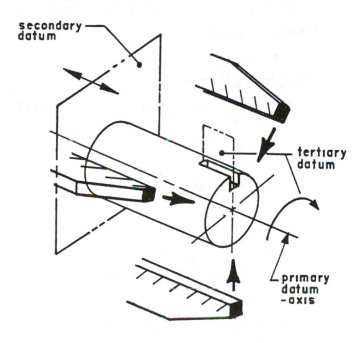

Figure 3-2 Axis as primary datum.

DATUM TERMS

Datum terms are shown in Figure 3-3. *Theoretic datums* are perfect, *simulated datums* exist in processing or verification equipment and simulate theory, *datum features* contain processing errors, and *temporary datums* are for manufacturing and/or inspection only and may or may not exist in the part's finished state.

When we refer to parts, features, and datums, it is important for all parties to use the same terms. Generally, when referring to datums, we mean the *theoretical* datum features and axes involved.

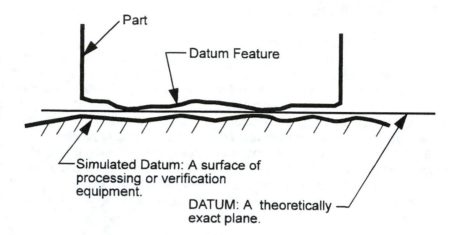

Datum

Theoretically perfect points, lines, or planes

Simulated Datum (Datum Simulators)

Surfaces and axes of processing or inspection equipment

Datum Feature

Actual part feature surfaces

Temporary Datum

Introduced for processing or inspection purposes (may be removed)

Figure 3-3   Datum terms.

DATUM SELECTION

Now that we are familiar with datum theory, datum frames, and datum terms, we must address the process of selecting datums for the design intent required. This process is generally based on the following:

Function and functional relationships: This is the most critical issue, impacting not only the design but the manufacturing and quality plans as well.

Reality: Datums should be real, identifiable, and verifiable. Imaginary features, points in space, or features that are impossible (or at least difficult) to locate are acceptable in a lab or mathematical environment, but are difficult for the manufacturing world to deal with. Equally difficult are datum features that are inaccessible or hidden within the product.

Accuracy: Datums should be accurate and should offer the best repeatability, because datum error can impact assessment and verification of other features. If a datum feature is within its specification control (flatness for example), it becomes theoretically perfect (or zero) for measurement of other features.

DATUM FEATURE SYMBOLS

The datum feature symbols of the ISO have been adopted by the United States. The symbols consists of a capital letter enclosed in a square box with a leader line extending from the box to the concerned datum feature and terminating with a triangle. Datum feature symbols illustrated in Figure 3-4 are applied as follows:

(a) Symbols are placed on the outline of a feature surface or extension line of the feature outline (but clearly separated from the dimension line) when the datum feature is represented by the feature surface or extension line.

(b) Symbols are placed on an extension of the dimension line of a size feature when the datum is the feature axis or centerplane. If space is limited, one of the arrowheads may be replaced by the datum triangle symbol.

(c) Symbols are placed on the outline of a cylindrical feature surface or an extension line of the feature outline, separated from the size dimension, when the datum is the feature axis.

(d) Symbols are placed below and attached to the feature control frame when the feature or group of features controlled is the datum feature axis or centerplane.

(e) Symbols are placed on the drawing planes established by datum target points on irregular or complex feature surfaces. (Refer to Figure 3-37 and Chapter 5.)

DATUM CENTERPLANES (AXES)

It is generally considered unwise to identify centerlines of features, or centerlines between features, as datums.See Figure 3-5. Three holes in line create a problem, in that only two of the holes can ever be aligned perfectly. The third (any hole) will have some displacement error.

Hole centerlines at 90 degrees also create potential errors in that datums are mutually 90 degrees by definition, which would be virtually impossible to re-create in a manufacturing environment.

Similarly, shafts with multiple diameters create a dilemma in that one must determine which features really constitute the datum axis. As shown in Figure 3-5, more than one feature may be sharing the illustrated axis or centerplane, and it may not be clear for which feature(s) the datum identifier was intended. Drilled and counterbored holes and/or holes hidden behind other holes are also examples.

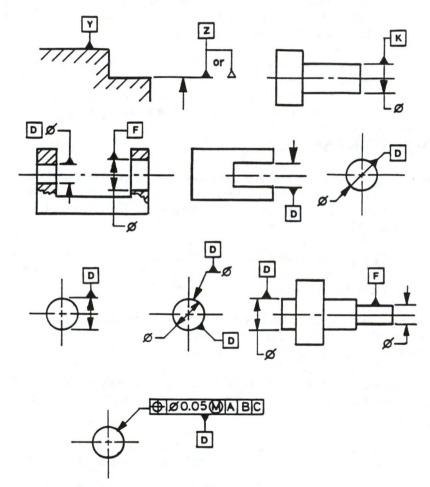

Figure 3-4 Datum symbols.

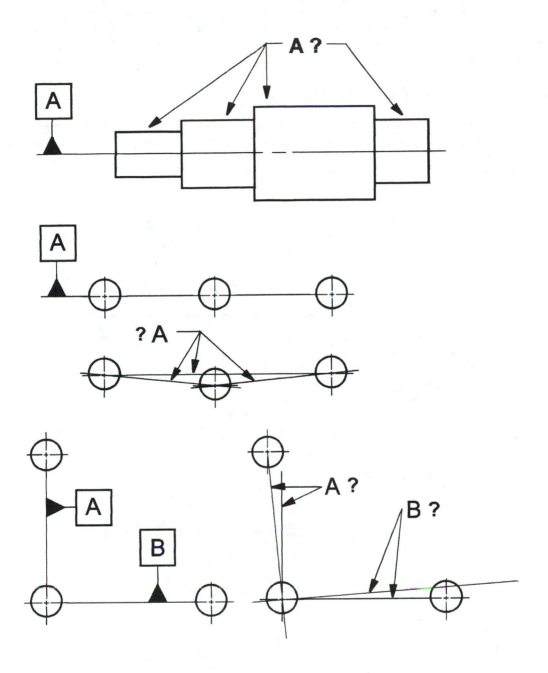

Figure 3-5 Datum symbols applied to axes.

DATUM FEATURE WITH FORM TOLERANCE

Primary datum features will generally have some sort of form control applied. For flat features, flatness or straightness of surface elements may be used, whereas with cylindrical features, cylindricity or circularity is common. For more complex shapes, a profile of line elements or of a surface may be used. Figure 3-6 illustrates a controlled cylinder that is used as datum feature A.

SECONDARY/TERTIARY DATUM RULE

When a feature of size (with MMC applied) is used as a *secondary or tertiary datum*, it must be simulated at its virtual condition if functional gaging is used. See Figure 3-7a. To determine the size of the secondary datum pin B, we must allow for the possibility for datum feature B being out of perpendicular with primary surface A. The same rule applies to tertiary datum slot C.

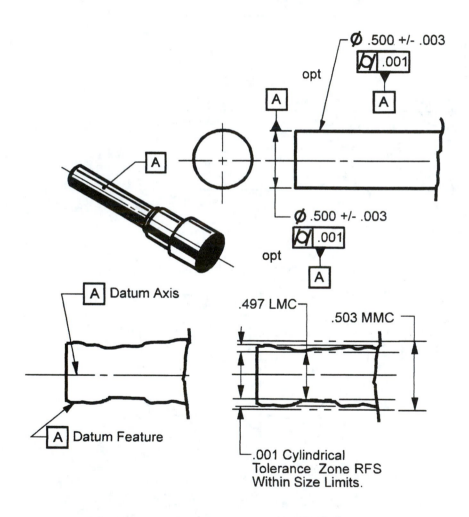

Figure 3-6   Datum feature with form tolerance.

66

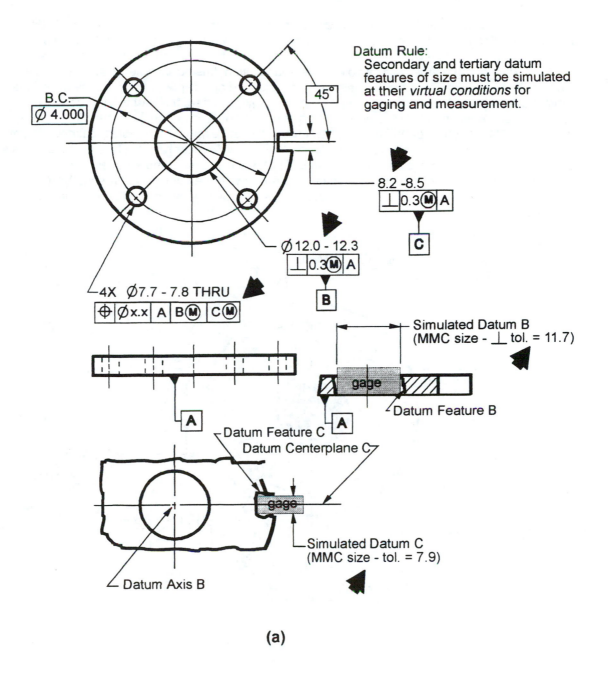

Datum Rule:
Secondary and tertiary datum features of size must be simulated at their *virtual conditions* for gaging and measurement.

B.C.
Ø 4.000

45°

8.2 -8.5
⊥ 0.3Ⓜ A
C

Ø 12.0 - 12.3
⊥ 0.3Ⓜ A
B

4X Ø7.7 - 7.8 THRU
⊕ Øx.x A BⓂ CⓂ

A

Simulated Datum B
(MMC size - ⊥ tol. = 11.7)

gage

Datum Feature B

Datum Feature C
A
Datum Centerplane C

Datum Feature C

gage

Simulated Datum C
(MMC size - tol. = 7.9)

Datum Axis B

(a)

Figure 3-7 (a) Secondary/tertiary datum rule;
(b) datums MMC and bonus tolerance.

67

DATUM FEATURES OF SIZE AND BONUS TOLERANCE

Figure 3-7b illustrates the feature holes controlled MMC, relative to datum surface A, datum diameter B (MMC), and datum slot C (MMC). The use of MMC for datums B and C implies a potential bonus tolerance available if the datum features depart from their MMC conditions. On occasion, depending on the feature hole's size and location, this bonus tolerance may not be useable. If the feature holes are diametrically opposite and are displaced in opposite directions, within the allowed tolerance zones, and if fixed pin gaging is used, the part will be locked in place. Any decrease in datum size will have no effect on any added bonus tolerance, as none is possible. If CMM or other open setup measurement is used, evaluation of hole locations, datum size, and bonus data must be carefully performed to achieve similar results. This condition exists for other figures in the text, as well as in ASME Y14.5 standard, when hole patterns are distributed about a secondary datum feature of size axis with the controls expressed MMC for both features and datums. ASME Y14.5M avoids this controversy by saying,"bonus tolerance may be available."

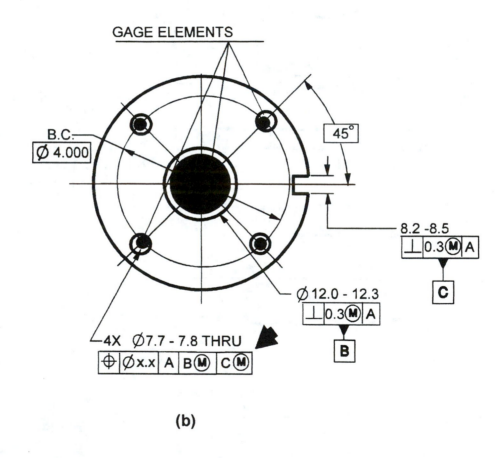

(b)

Figure 3-7   (continued)

DATUM ORDER (PRECEDENCE)

Design intent can be altered by changing the order of the datums, as illustrated by Figure 3-8. This shows the fundamentally different designs resulting from changing the order of the primary datum. Figure 3-8a shows diameter A as the primary datum RFS. To evaluate or measure, we locate on primary datum A, with any error being reflected to the secondary datum surface B. The holes must be located from and oriented to (parallel) primarily datum A. Figure 3-8b shows datum surface B as primary, with any error reflected to secondary datum A. The holes must be perpendicular to primary datum B and located from secondary datum A. Secondary datum A is the minimum cylinder contacting diameter A that is perpendicular to datum B. Figure 3-8c shows surface B as the primary datum again, with diameter A as the secondary datum MMC. This control callout will allow a functional receiver gage to be used. The gage would have a hole that simulates datum diameter A at MMC size (16.0), which is perpendicular to gage datum surface B within gage maker's tolerance. From this example, it is clear that the order of the datum precedence is as important as the identification of the datum features themselves.

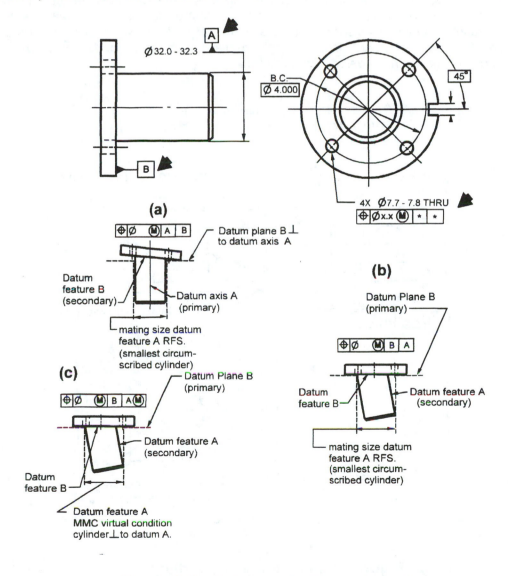

Figure 3-8   Datum order (precedence).

DATUM FRAMEWORK

Figure 3-9 illustrates examples of datum frameworks along with possible gage designs that have been developed from the functional relationships and control callout specified. If the datum requirements were the part edges, as in Figure 3-9a, the fixture and measurement process is fairly simple, and straightforward. When the secondary datum is a feature axis, the tertiary datum serves to stop the rotation, as in Figure 3-9b, and the evaluation process becomes more involved.

TERTIARY DATUMS

If the primary and secondary datums leave a *rotational degree of freedom*, the tertiary datum is basically oriented (in some cases also located) to the higher order of precedence datums. This relationship is shown in Figures 3-10 and 3-11. Further, if the datum framework is established by cylindrical features, not in the same plane, only *two* datums are required to complete the control framework. The intersection of the perpendicular planes that create the axes of datum features may be used in combination, as shown in Figure 3-11. The gage pins will locate and stop all movement of the part in the gage, thus "locking" it in place. In the paragraphs that follow, *datum frames and secondary/tertiary datums* are explored.

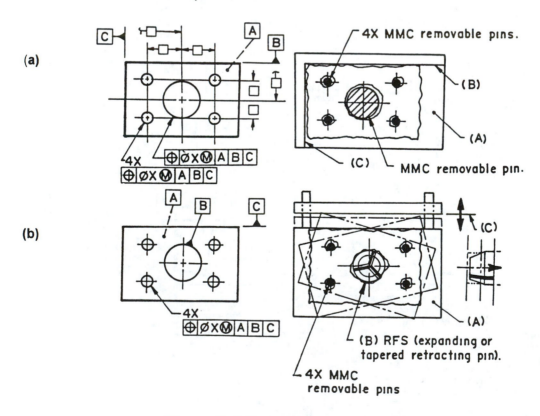

Figure 3-9 Secondary and tertiary
datum features.

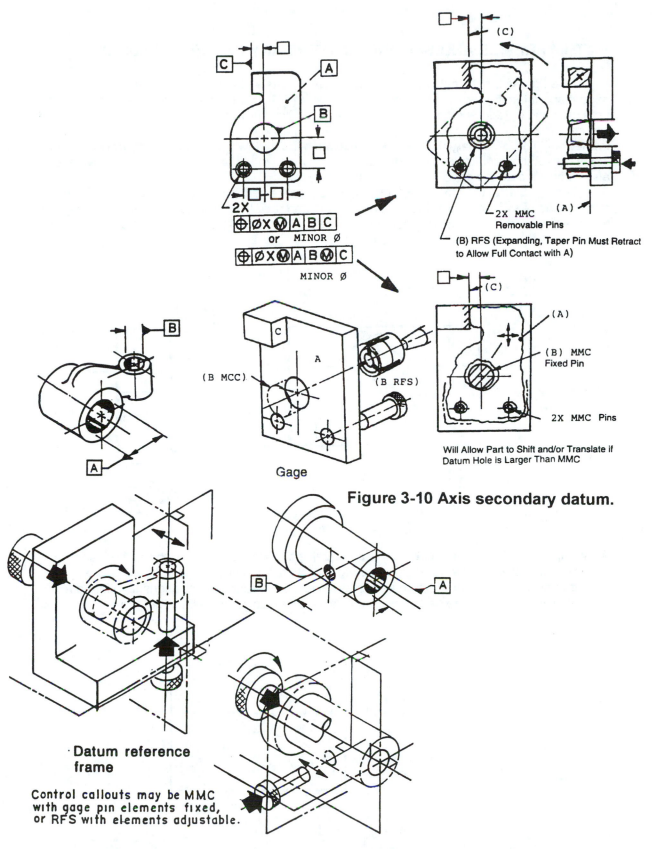

2X

⊕ ⌀X Ⓜ A B C
or   MINOR ⌀

⊕ ⌀X Ⓜ A B Ⓜ C
MINOR ⌀

2X MMC
Removable Pins

(B) RFS (Expanding, Taper Pin Must Retract
to Allow Full Contact with A)

(B MCC)

(B RFS)

Gage

(A)

(B) MMC
Fixed Pin

2X MMC Pins

Will Allow Part to Shift and/or Translate if
Datum Hole is Larger Than MMC

Figure 3-10 Axis secondary datum.

·Datum reference
 frame

Control callouts may be MMC
with gage pin elements fixed,
or RFS with elements adjustable.

Figure 3-11 Datum framework - two cylindrical features.

71

DATUM FRAMES AND SECONDARY/TERTIARY DATUM FEATURES

In the figures that follow, the primary datum feature is a plane surface, whereas the secondary datum feature is a hole or shaft diameter RFS. The secondary gage element may be an expanding or taper element that is retractable to allow full contact of the part with the primary datum surface. The primary datum establishes the relationship of the simulated datum framework. The secondary datum simulator locates the part, while the tertiary datum simulator stops the rotation and locks the part in the framework. In these examples, the datum frame is also referred to as the *gage* or *fixture*.

In Figure 3-12a, the tertiary datum is a plane surface and must be equalized about a vertical centerplane of symmetry. It serves to stabilize the part, and the simulator is relieved so that the fixture will accept a part with an imperfect (warped or curved) tertiary feature surface. It may be a *rocker*.

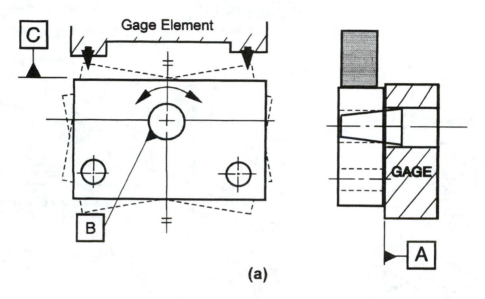

(a)

In Figure 3-12b, the tertiary datum is an equalized centerplane established by expanding buttons, which create isolated (in 360 degrees) points of contact with the tertiary feature slot (a feature of size). The tertiary feature simulator will establish alignment relationship aligned (or parallel) with the axis of the secondary datum simulator to lock the part in the fixture.

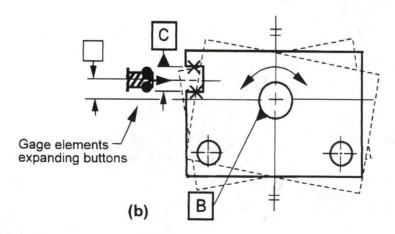

(b)

Figure 3-12

72

In Figure 3-12c, the tertiary datum simulator is a isolated point contact on a plane surface that stops rotation. The tertiary simulator must be located and oriented with respect to the higher precedence datum simulators.

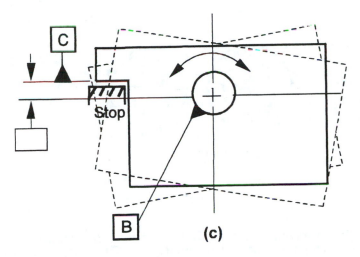

(c)

In Figure 3-12d, the tertiary datum feature is a feature of size, which may or may not be equalized or centered about the horizontal centerplane. The tertiary datum feature serves to stabilize the part, and gage elements may be relieved to avoid rocking. If location and alignment of datum simulator C with datum simulator B are maintained, it may not be possible to also contact both sides of datum feature C simultaneously. This example then becomes similar to Figure 3-12a.

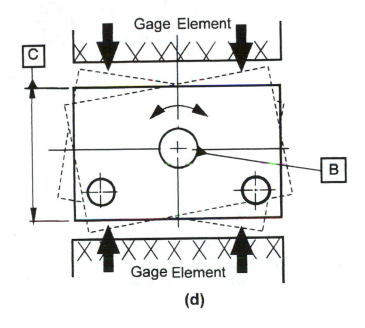

(d)

Figure 3-12   (continued)

In Figure 3-12e, the workpiece is symmetrical about the vertical centerplane. The tertiary datum feature is a plane surface (two surfaces) used to stabilize the workpiece.Conditions are similar to Figure 3-12a.

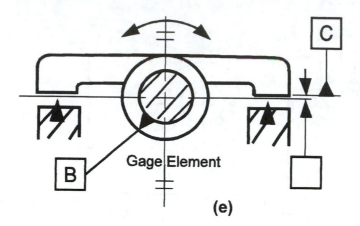

(e)

Figure 3-12f is similar to Figure 3-12e but with the tertiary datum feature as a single surface (point) of contact to stop the rotation. This surface or point must be located and oriented to the primary and secondary datum so as to isolate the workpiece in the fixture.

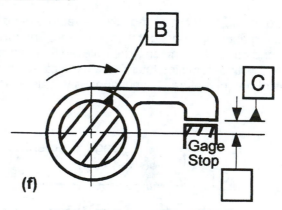

(f)

In Figure 3-12(g), the secondary datum feature is the shaft (pin) diameter, whereas the tertiary datum feature is a slot (feature of size). The conditions and requirements are the same as for Figure 3-12b.

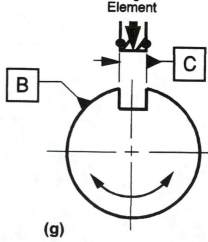

(g)

Figure 3-12 (continued)

In Figure 3-12h, the tertiary datum feature is a feature of size (hole) used to stop rotation. The same conditions exist as for Figures 3-12b and 3-12g.

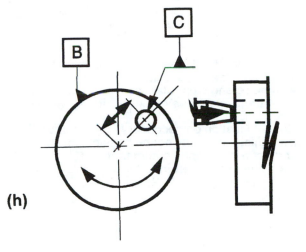

(h)

In Figure 3-12i, the tertiary datum feature is a feature of size (slot) used to stop rotation. The same conditions exist as for Figures 3-12b,g, and h.

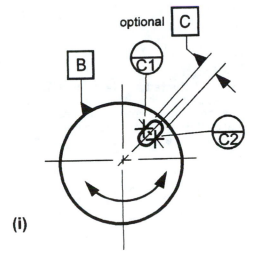

(i)

In Figure 3-12j, the tertiary datum feature is a surface (point) isolated in 360 degrees to stop the rotation. The same conditions exist as for Figures 3-12c and 3-12f.

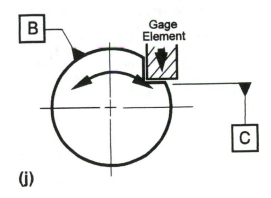

(j)

Figure 3-12 (continued)

In Figure 3-12k, the tertiary datum feature is a slotted feature of size used to stabilize the workpiece and stop rotation. The feature may be at the centerplane or axis, or it may be offset.

Note: If location on secondary datum feature B is maintained, it may not be possible to also contact both sides of datum feature C simultaneously; therefore, this example is similar to Figures 3-12a and 3-12d.

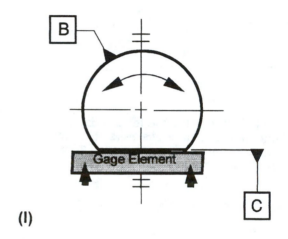

(k)

In Figure 3-12l, the tertiary datum feature is a surface used to stabilize the workpiece and stop rotation. Gage elements may be relieved to allow for any bowed condition of feature C (rocker). The same conditions exist as for Figure 3-12a.

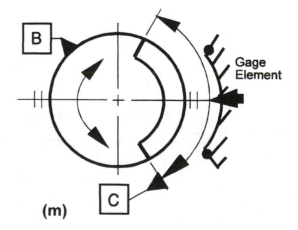

(l)

In Figure 3-12m, the tertiary datum feature is a circular notch in the datum B surface. It is considered a feature of size, having a center-plane, which requires alignment with the secondary datum feature axis, and orientation (perpendicularity) to the primary feature. The tertiary datum stops rotation. The same conditions exist as for Figures 3-12g, h, i, and k.

(m)

Figure 3-12   (continued)

In Figure 3-12n, the tertiary datum feature is a feature of size (hole) used to stop rotation. The same conditions exist as for Figures 3-12g, h, i, k, and m. When two cylindrical features that do not lie in the same plane are used as datum features, the two are sufficient to complete the datum reference framework.

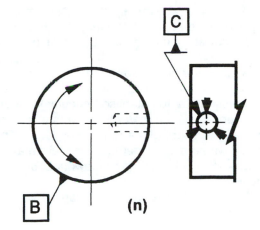

(n)

In Figure 3-12o, two simultaneous surfaces serve as the tertiary datum feature(s) and are used to stabilize the workpiece while stopping rotation. They must be located and oriented to the secondary and primary datum features. The same conditions exist as for Figures 3-12a, e, and l, except that no rocking would occur.

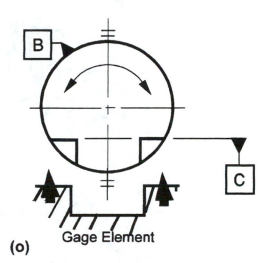

Gage Element

(o)

Figure 3-12   (continued)

In conclusion,

A *plane surface tertiary datum* simulator seems to serve as a control of radial orientation only, when the surface is equally distributed about a centerplane of symmetry, because the feature surface must be movable so as to stabilize perpendicular to the centerline or centerplane of symmetry.

A *plane surface tertiary datum* simulator can be used as a stop to control radial orientation about a secondary feature axis, but it must be located and oriented from the secondary datum feature simulator to establish the complete datum framework.

A tertiary *datum feature of size* that is located relative to a centerline or centerplane of symmetry will control the radial orientation. The tertiary datum feature of size may also be aligned with the secondary datum feature axis, or separated by a basic dimension, when the two are locked together in the framework and/or when it is illustrated they share an inseparable function.

Figure 3-12p is one possible application of a design with a primary datum axis and secondary datum target points, similar to the method referred to in Figure 3-12i.

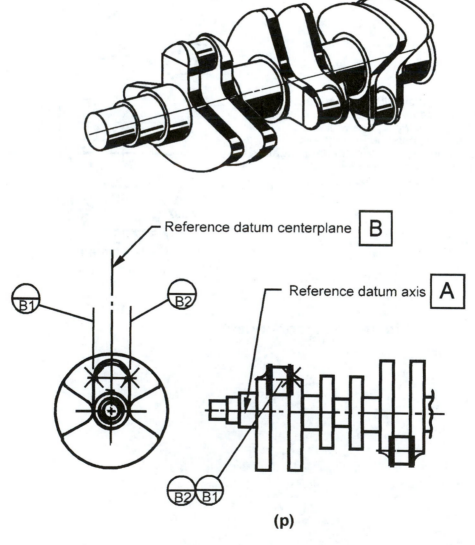

(p)

Figure 3-12   (continued)

STEPPED DATUM FEATURE

Three points will establish a plane. Figure 3-13 illustrates how any given point may be offset by a basic amount, and how, collectively, the three points establish datum A. This condition is common to many parts and is referred to as a *Stepped Datum*. The 25 mm basic dimension is for the basic separation and also for tool/gage tolerance application. Note the use of the *dimension origin symbol* indicating the exact location of the tolerance zone.

Target point symbols have also been introduced to identify the three precise locations for the points that establish datum A. Datum targets are discussed later in this section.

PARTIAL DATUM FEATURES

Part features that are very long or otherwise inappropriate to use in their entirety as a datum reference are often encountered. When this occurs, a portion of the feature may be selected as a *partial datum feature*. These limited lengths or areas may be identified and specified by the combined use of basic dimensions and target points. The use of datum symbols and identifying portions of the object lines is also acceptable. See Figure 3-14.

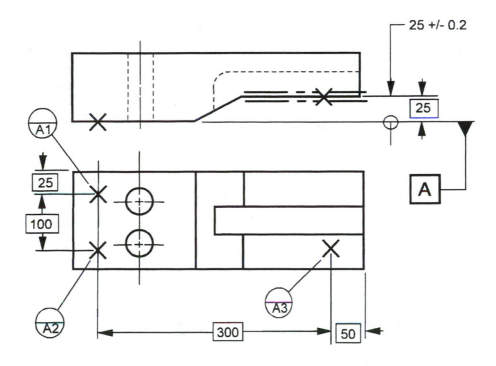

Figure 3-13   Stepped datum feature.

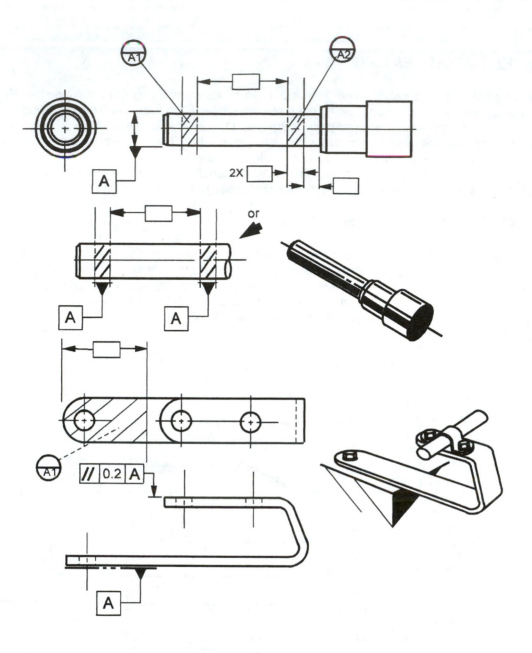

Figure 3-14 Partial datum.

COPLANAR AND COAXIAL DATUM FEATURES

Coplanar or coaxial datums exist when two features are used simultaneously to create a single datum reference (see Figure 3-15). In this figure, the datum symbols appear in the same box of the control frame, but they are separated by a dash (A-B). This callout indicates that the two features are used as one. Further, it is acceptable to mix combinations of datum callouts and datum target callouts in the feature control frame. In the case of a shaft, the entire feature may be used, or selected locations on the feature, located with basic dimensions, and identified with datum target symbols, may be used. The target system may be needed if precise process and measurement repeatability is mandatory, or with multiple plants or suppliers involved in the process. The enlarged view of Figure 3-15 illustrates how the use of the entire feature and the use of target locations on a feature may give varied results. For critical designs, this issue should be resolved in the review process with manufacturing and quality control prior to the release of the design to manufacturing agencies. See also Figures 4-8 and 5-8.

80

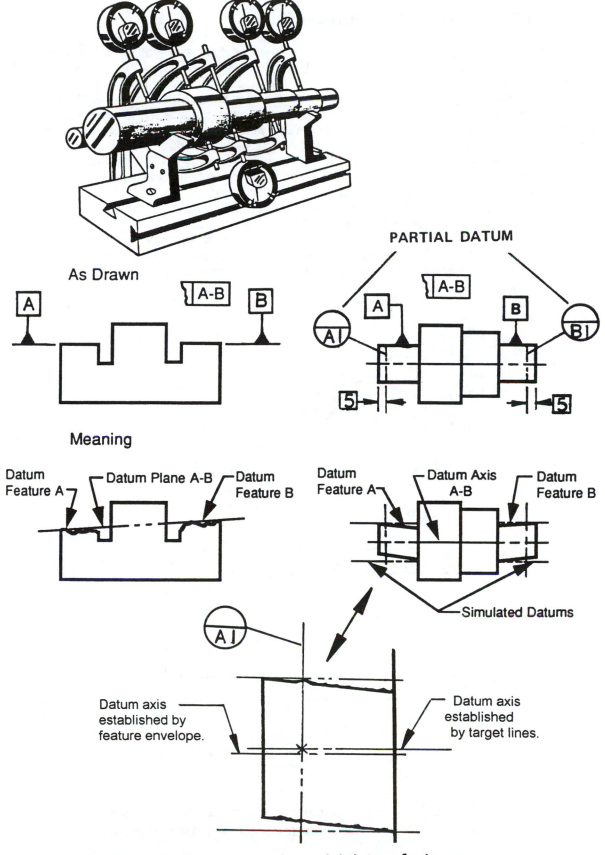

Figure 3-15 Coplanar and coaxial datum features.

HOLE PATTERNS AS DATUM FEATURES

In some circumstances, patterns of holes may be used as datums. Not all parts are stable or designed such that datums are easily identifiable or practical to use. Gaskets, moldings, and stampings are examples. When no functional datum features are available, consider the pattern of mounting holes as a possible option. Figure 3-16 illustrates a part with many holes. This type of part may be a cover plate or sump pan. By using the mounting surface as the primary mounting surface and the hole pattern MMC as the secondary/tertiary datum, a receiver gage can be an option for acceptance of parts. The part could slip over the pattern of gage pins, and other features, including the inner and outer contours could be evaluated. This technique is popular on high-volume parts where acceptance is the primary concern. Basic dimensions would be required for the hole locations as well as the contour dimensions.

Hole patterns may be used as datums when necessary to locate (center) another feature (large hole) in the pattern, as shown by Figure 3-17a. This technique ensures that the pattern is centered in the part, symmetrically located from feature centerplanes, and with the large hole centered within the pattern of four smaller holes when the datum holes and feature hole all are MMC. The center of the pattern of holes is determined by the size and location of all holes simultaneously. See Figure 3-17b.

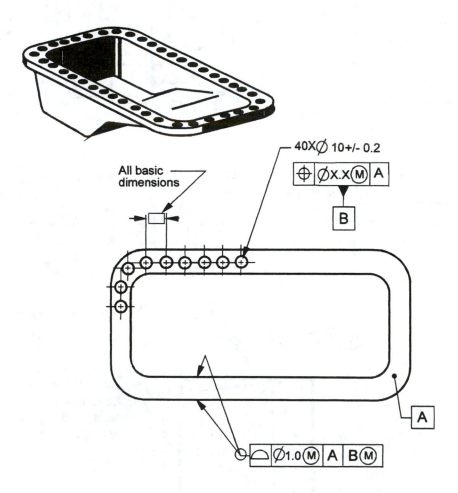

Figure 3-16   Hole pattern datum: multiple features of size as secondary/tertiary datum.

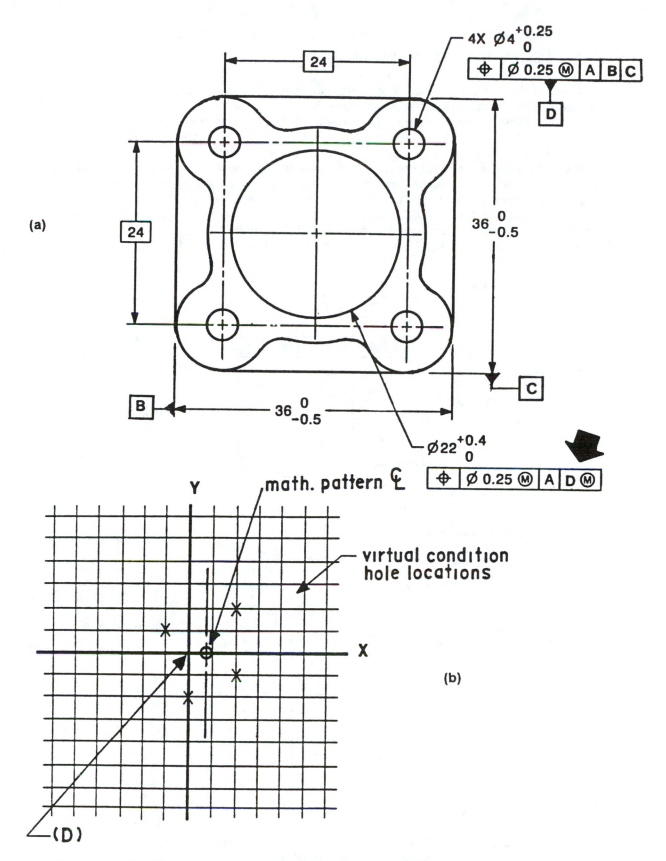

Figure 3-17 Hole pattern datum and mathematical pattern center.

FEATURE DATUM CENTERPLANES

Rectangular or square feature edges are usually envisioned as being perfect at 90 degrees from each other. In manufacturing a part, however, it is rare that a set of features will be perfect. Determining those feature centerplanes can be difficult in these cases. Figure 3-18 illustrates imperfect parts, using the datum centerplane principle. The figure shows the construction of sets of parallel planes, with the centerplane between these planes as datum centerplanes. Sets of parallel planes are established from the extremes of the part edges. These planes are mutually 90 degrees by definition. The centerplanes of this constructed geometry are centerplanes B and C. If the MMC principle is applied to the feature callouts, it would be practical to construct a receiver box gage, 6.050 X 4.050 in size, with four .475 diameter gage pins basically located from the datum centerplanes. If applied on an RFS basis, for both holes and datum features, measurements would be taken from the centerplanes of the constructed geometry, RFS.

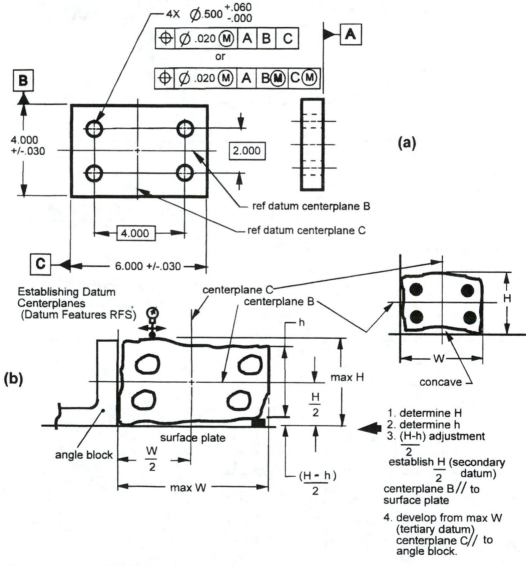

Figure 3-18   Position applications: symmetry and datum centerplanes.

84

If the part were found to be skew as in Figure 3-18b, the maximum width at one end could be found, and by use of shims or mathematically adjustment, the other end could be equalized to construct a framework for the mutually perpendicular planes (datum framework). Other techniques are possible, including the use of Coordinate Measuring Machines (CMMs) to mathematically simulate the geometric model, and associated centerplanes.

DATUM REFERENCE FRAMES THROUGH TWO HOLES

Datum reference frames are often established by a surface (primary) and two holes serving as secondary and/or tertiary. Figures 3-19 illustrates the concepts. Holes are features of size, the use of MMC, RFS, or LMC must be considered in the design controls.

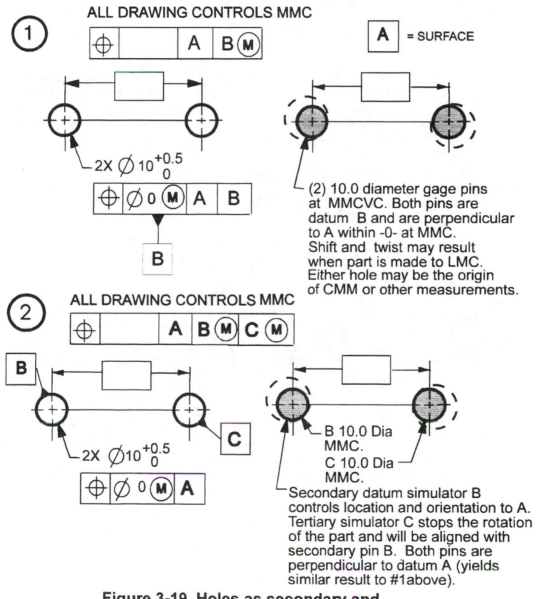

Figure 3-19 Holes as secondary and tertiary datums.

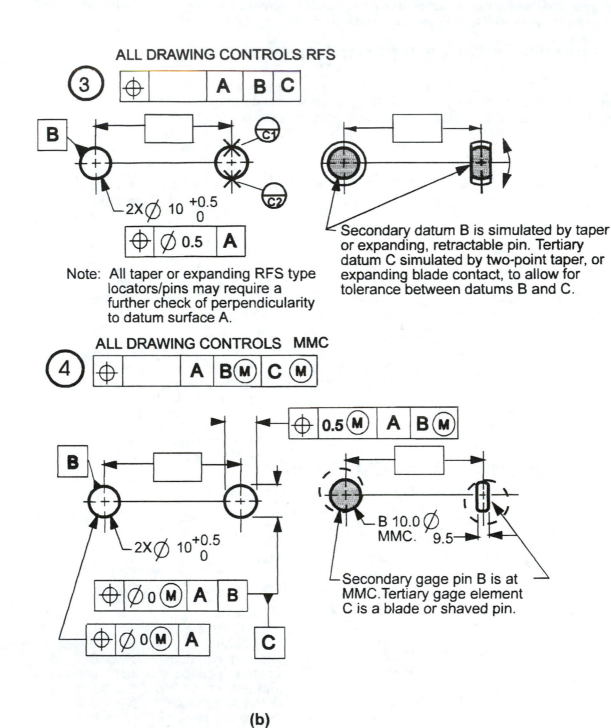

ALL DRAWING CONTROLS RFS

③ ⊕ | | A | B | C

B

2X Ø 10 +0.5 / 0

⊕ | Ø 0.5 | A

C1

C2

Note: All taper or expanding RFS type locators/pins may require a further check of perpendicularity to datum surface A.

Secondary datum B is simulated by taper or expanding, retractable pin. Tertiary datum C simulated by two-point taper, or expanding blade contact, to allow for tolerance between datums B and C.

ALL DRAWING CONTROLS MMC

④ ⊕ | | A | B Ⓜ | C Ⓜ

B

2X Ø 10 +0.5 / 0

⊕ | 0.5 Ⓜ | A | B Ⓜ

⊕ | Ø 0 Ⓜ | A | B

⊕ | Ø 0 Ⓜ | A

C

B 10.0 Ø MMC. 9.5

Secondary gage pin B is at MMC. Tertiary gage element C is a blade or shaved pin.

(b)

Figure 3-19   (continued)

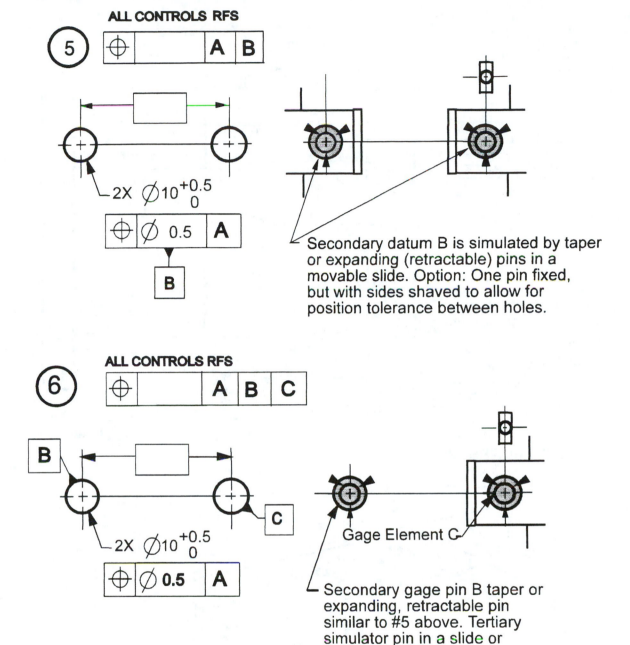

⑤ ⊕ | | A | B

2X ⌀10 +0.5 / 0

⊕ | ⌀ 0.5 | A

B

Secondary datum B is simulated by taper or expanding (retractable) pins in a movable slide. Option: One pin fixed, but with sides shaved to allow for position tolerance between holes.

⑥ ⊕ | | A | B | C

B

C

2X ⌀10 +0.5 / 0

⊕ | ⌀ 0.5 | A

Gage Element C

Secondary gage pin B taper or expanding, retractable pin similar to #5 above. Tertiary simulator pin in a slide or with shaved sides.

(c)

Figure 3-19   (continued)

Figure 3-20a illustrates a six-hole pattern design with the datum holes at MMC (or RFS). The feature holes are shown at MMC. In the explanations of Figure 3-20b, note that the datum holes pins must allow for location tolerance of the datum holes, and if specified RFS, this allowance is reflected in flexibility of the gaging. This flexibility increases gaging expense. Further, it is important to consider other measurement devises, such as CMMs. The quality level of products accepted must be the same, regardless of the measurement technique used.

Additional positional tolerancing and gaging exercises are covered later in the text in Chapters 7 and 8.

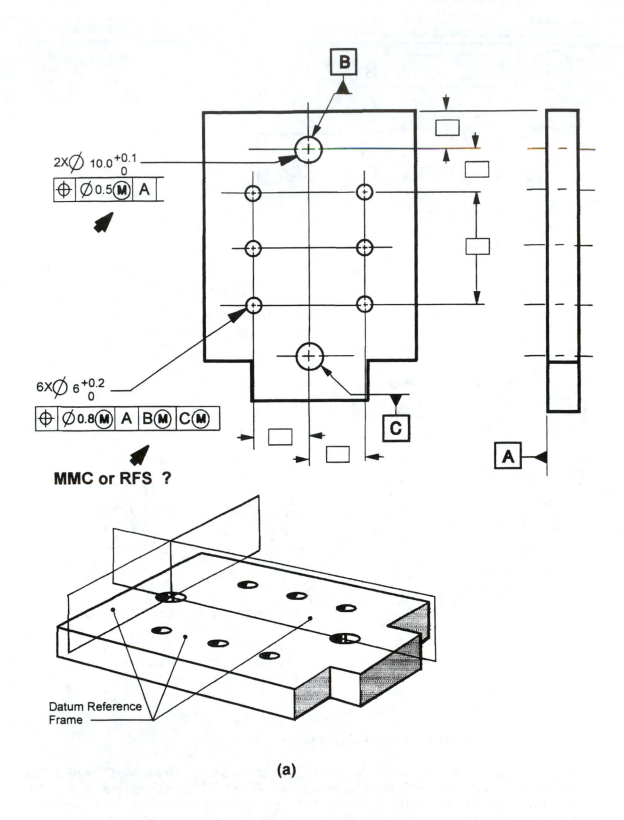

2X⌀ 10.0 +0.1/0

| ⊕ | ⌀0.5 Ⓜ | A |

2X⌀ 6 +0.2/0

| ⊕ | ⌀0.8Ⓜ | A | BⓂ | CⓂ |

MMC or RFS ?

B

C

A

Datum Reference
Frame

(a)

Figure 3-20   Datum reference framework: thru two holes.

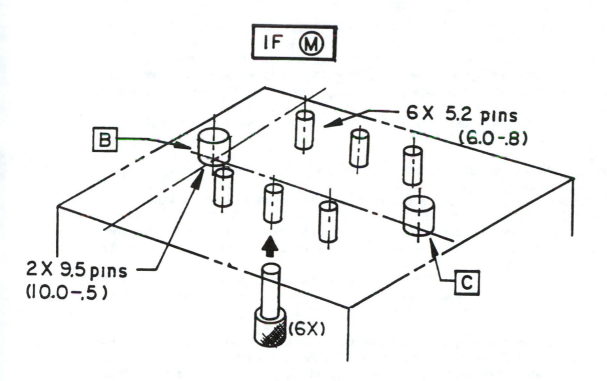

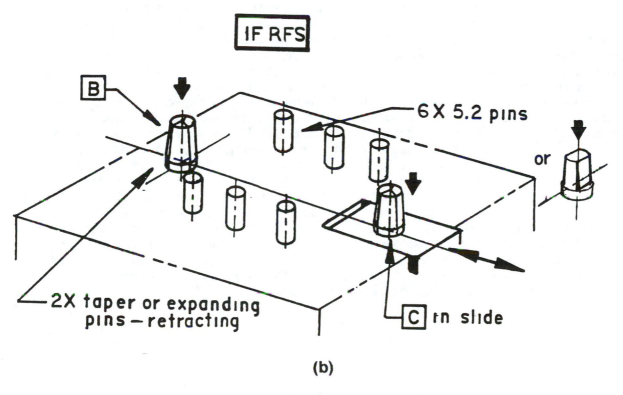

GAGES

Individual datum axes are established at the true position of each datum hole (at virtual condition). When mounted on the primary datum surface, the holes axes serve as the secondary and tertiary reference frame

(b)

Figure 3-20   (continued)

Figures 3-21 through 3-28 illustrate the effects of material condition modifiers on feature and datum controls. Study each figure and examine the impact of design controls on gage designs. Note that as the datums and modifying controls change, new and varied design requirements emerge.

Figure 3-21 shows the datum holes at MMC located from the datum framework A,F,D. The six feature holes are at MMC and are located from A, B at MMC, and C at MMC. See gage Figure 3-22.

Figure 3-23 shows the added requirement of the datum holes at MMC, perpendicular to surface A. See gage Figure 3-25a and b.

In Figure 3-24 the datum holes are located from the framework A, F, and D within zero tolerance MMC. See gage Figure 3-25c.

In Figure 3-26 the feature holes MMC are located from the datum holes RFS. See gage Figure 3-27. Figure 3-26 also shows the datum holes RFS located from the part framework A, F, D. See gage Figure 3-27.

In Figure 3-28 the datum holes (of equal importance) at MMC are located from framework A, F, and D. See Figure 3-28a. The secondary datum axis B for the location of the smaller holes is established at the centerplanes through both datum B holes. See gage Figure 3-28b.

Figure 3-21 shows that the two 10 mm holes are the secondary and tertiary datums. The control callout shows both holes are in true position relative to the datum framework A, F, and D within a 0.5 dia. tolerance zone when both holes are at MMC. Figure 3-22a graphically illustrates a possible functional gage for this control.

Figure 3-21 also illustrates that the six 6 mm feature holes are to be located from surface A and the two datum holes when the datum holes are at MMC. The tolerance for the six holes is a 0.8 mm cylinder. Figure 3-22b illustrates a possible functional gage for this control. Functional gaging is an option to consider because the feature holes as well as the datum holes are specified at MMC.

Note in Figure 3-22b that six loose pins have been used for gaging the small holes. Fixed pins could have been used. If the part is placed on the gage, with full contact on the primary datum surface, loose pins can evaluate individual holes acceptance. This technique helps give insight about which holes are out of location, what fixture or spindle rework or repair is required, or if replacement is necessary. With six fixed pins, the gage will tell us if the part is good or bad, but it will be difficult to determine why, and to determine which holes are out of spec.

Figure 3-23 illustrates a further control for the datum holes of perpendicularity y to surface A within a 0.2 diameter tolerance. The datum holes are to be located within a 0.5 diameter tolerance from framework A, F, and D at their MMC size limit. Figure 3-25a and 3-25b illustrates the impact of this added control on gage designs. Both the secondary and tertiary datum holes must be perpendicular to surface A within a 0.2 diameter tolerance zone (9.8 dia gage pins). Because the secondary gage pin has not allowed for any location tolerance variation, this potential 0.5 diameter tolerance must be reflected in the tertiary datum. See Figure 3-25a.

Because each horizontal entry in the control frame is a self-contained, separated specification, it may be verified independently of other controls. Figure 3-25b illustrates the use of two 9.8 diameter pins for the secondary and tertiary datum holes to verify for perpendicularity of 0.2. Because there may be some positional error between datum holes, the tertiary datum pin is placed in a movable slide to allow for possible error. The six 5.2 feature hole pins could be added to this gage as well, giving an optional gage design.

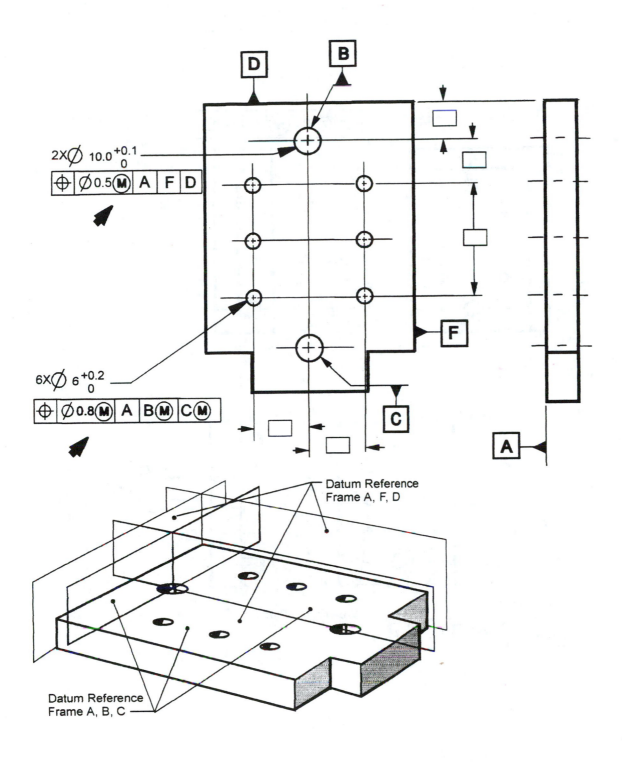

Figure 3-21 Datum reference frame: two holes.

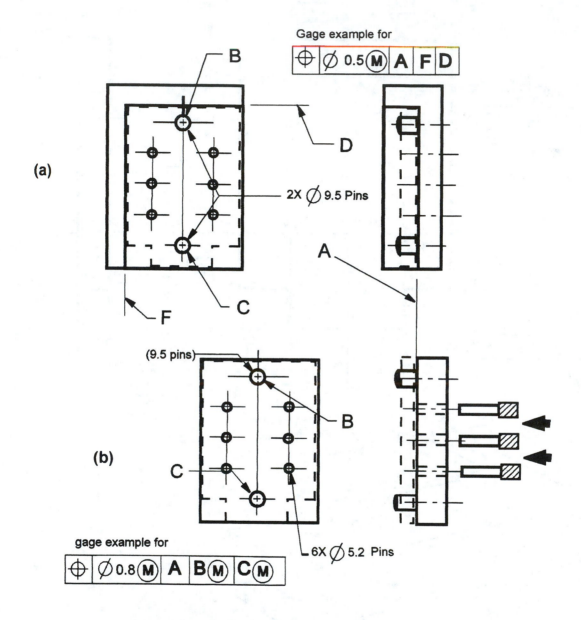

(a)

Gage example for

⊕ | ⌀ 0.5 Ⓜ | A | F | D

B

D

2X ⌀ 9.5 Pins

A

F

C

(b)

(9.5 pins)

B

C

6X ⌀ 5.2 Pins

gage example for

⊕ | ⌀ 0.8 Ⓜ | A | B Ⓜ | C Ⓜ

Figure 3-22 Explanation for Figure 3-21.

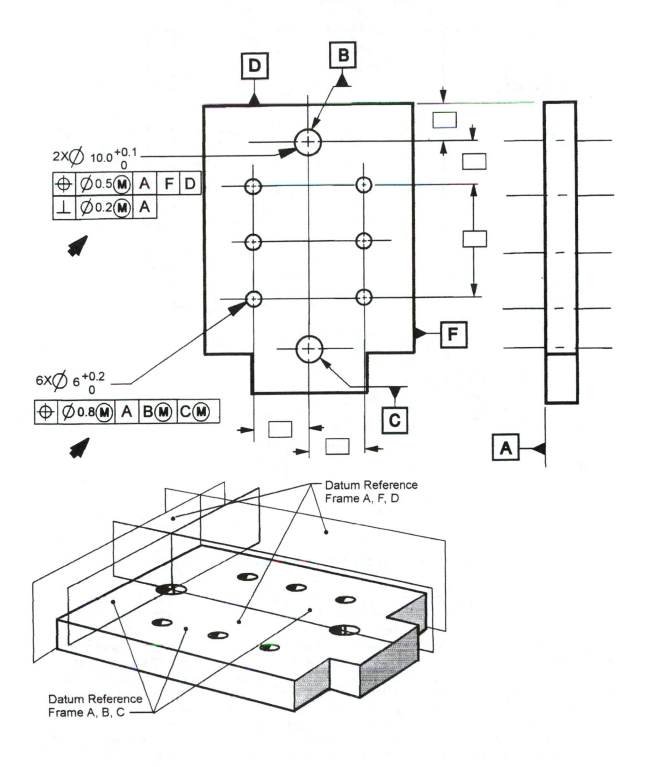

Figure 3-23 Datum reference frame: two holes.

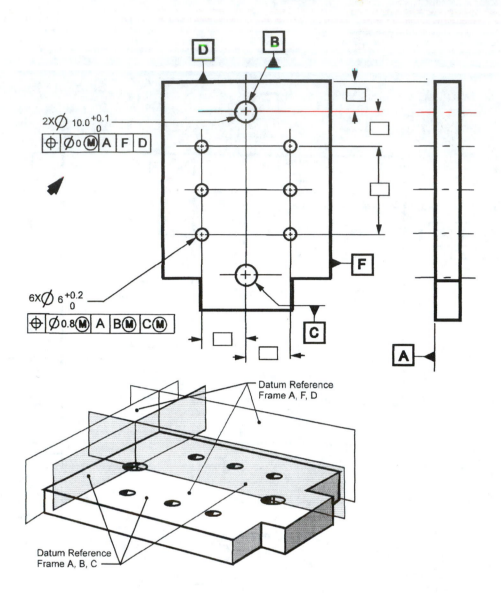

2X⌀ 10.0 +0.1/0

| ⊕ | ⌀0 Ⓜ | A | F | D |

6X⌀ 6 +0.2/0

| ⊕ | ⌀0.8 Ⓜ | A | BⓂ | CⓂ |

D   B

F

C

A

Datum Reference
Frame A, F, D

Datum Reference
Frame A, B, C

Figure 3-24   Datum reference frame: two holes.

Figure 3-24 illustrates that the two 10 mm datum holes are to be located from the datum framework within zero tolerance at MMC. Figure 3-25c shows a possible gage for this control.

Figure 3-26 illustrates two datum holes located within a 0.5 diameter tolerance zone RFS. Because the tolerance is expressed RFS, relative to datum framework A, F, and D, a functional gage is not practical. Measurements must be taken via open setup inspection, CMMs, or other variable measurement systems. See Figure 3-27. Figure 3-26 also illustrates the six 6 mm holes, at MMC, to be located from surface A and the two datum holes RFS. This control callout permits six 5.2 diameter gage pins for the MMC feature holes; however, because the datum holes are expressed RFS, some variables must be built into the gage to allow for possible tolerance variation relative to the datum holes. Figure 3-27 illustrates a technique of using tapered, retractable pins, allowing the part to make full surface contact with primary datum surface A, aligning on secondary datum hole B and tertiary datum hole C.

Because of possible location tolerance variations between datum holes, the pin for tertiary datum C should be in a movable slide. Although this approach comes close to evaluating the datum holes location to the datum framework A, B, and C, it is not exact in that it does not verify the perpendicularity of the datum holes to surface A. The tapered pins only contact the datum holes at surface A, not through the full depth of the holes.

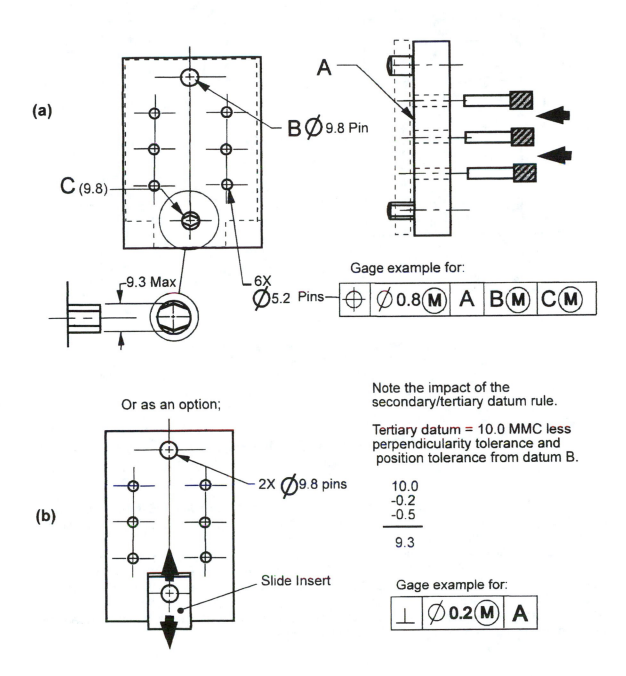

Note the impact of the secondary/tertiary datum rule.

Tertiary datum = 10.0 MMC less perpendicularity tolerance and position tolerance from datum B.

$$\begin{array}{r} 10.0 \\ -0.2 \\ \underline{-0.5} \\ 9.3 \end{array}$$

Figure 3-25 Explanation for Figure 3-23.

95

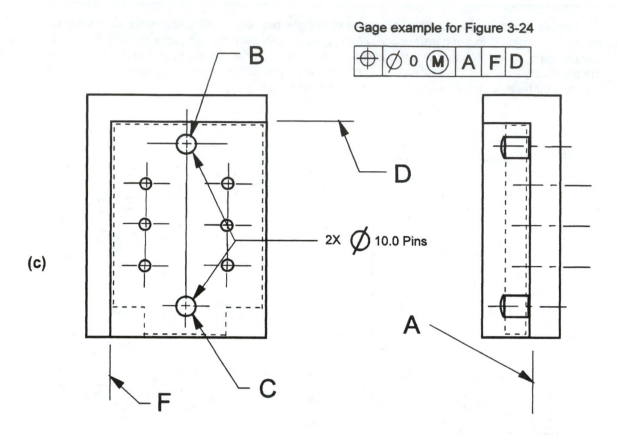

Gage example for Figure 3-24

⊕ | ⌀ 0 Ⓜ | A | F | D

(c)

2X ⌀ 10.0 Pins

Note: Zero position tolerancing is possible
only when applied for features at MMC.

Figure 3-25 (continued)

Figure 3-28a illustrates the datum holes (as a pair) as secondary datum B, MMC. No tertiary datum is expressed or needed. This condition is illustrated in Figure 3-28b. Neither datum hole has more design influence.

Figure 3-28a also shows the six small holes located from datum surface A and the two datum holes at their virtual condition MMC. The axis of the secondary datum B goes through both holes, no matter which is first; therefore, the three plane framework may be established without a tertiary datum expressed. The tertiary datum plane may be constructed through either hole. The datum holes may have been shown in the lower control frame RFS; however, adjustable/variable setup methods as shown in Figure 3-27 would be required.

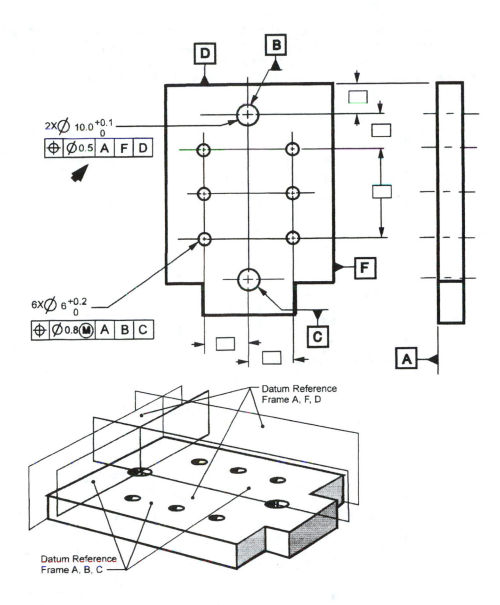

2X∅ 10.0 $^{+0.1}_{0}$

| ⊕ | ∅0.5 | A | F | D |

6X∅ 6 $^{+0.2}_{0}$

| ⊕ | ∅0.8Ⓜ | A | B | C |

Datum Reference
Frame A, F, D

Datum Reference
Frame A, B, C

Figure 3-26 Datum reference frame: two holes.

DATUM TARGETS

Datum targets are often used to specify points, lines, or areas of surfaces that are to be used to specifically control and clarify design requirements. Targets may be used on rough surfaces, such as cast or forged surfaces, or on finished surfaces. When used on rough surfaces, target points or areas qualify parts to ensure adequate machining stock and provide uniform contact points for fixturing operations, thus minimizing scrap, helping to ensure repeatability in the acceptance process, and clearly aiding communication with suppliers.

Because of inherent irregularities, the entire surface of some features cannot be effectively used to establish datums. Examples include nonplanar or uneven surfaces produced by the processes above and also by molding, weldments, stamping, etc. Thin section surfaces subject to bowing, warping, or other distortions, are also included. Combinations of datum features and target points/areas may be used.

97

Target points are also used on finished surfaces for precise controls to maintain measurement repeatability and to ensure interchangability and gage commonality due to multi-plant or supplier sourcing. It is important to note that more precise controls mean less flexibility for the manufacturing process in general. See Figure 3-29. Application of target points are shown in Figures 3-30 through 3-34.

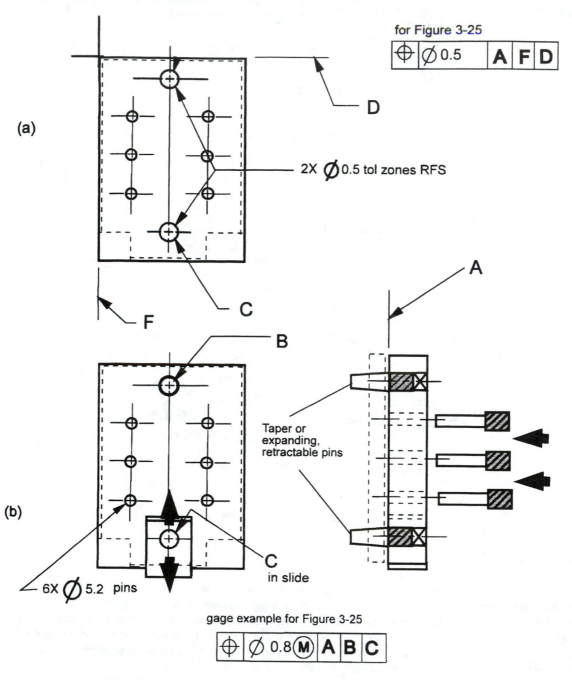

Figure 3-27 Explanation for Figure 3-26.

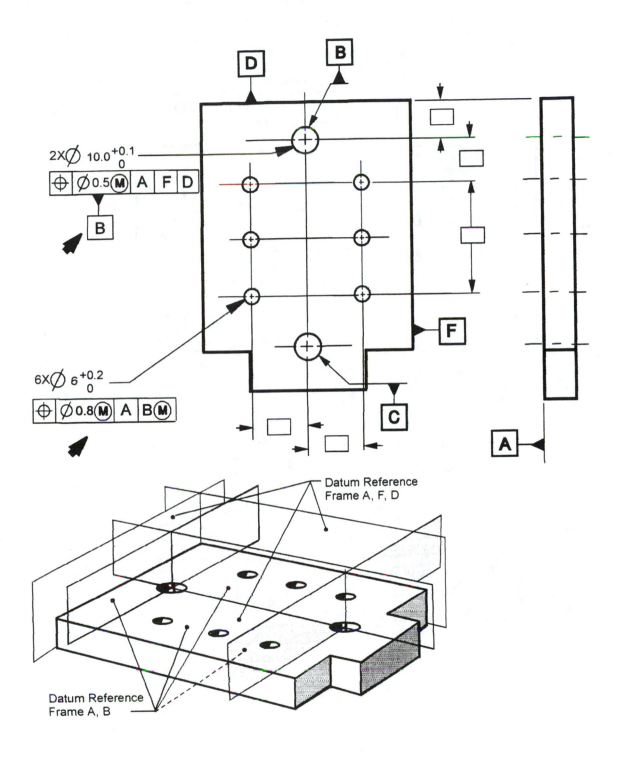

$2X \emptyset \ 10.0^{+0.1}_{\ 0}$

| ⊕ | $\emptyset 0.5$Ⓜ | A | F | D |

B

$6X \emptyset \ 6^{+0.2}_{\ 0}$

| ⊕ | $\emptyset 0.8$Ⓜ | A | BⓂ |

D   B

F

C

A

Datum Reference
Frame A, F, D

Datum Reference
Frame A, B

Figure 3-28 Datum reference frame: two holes.

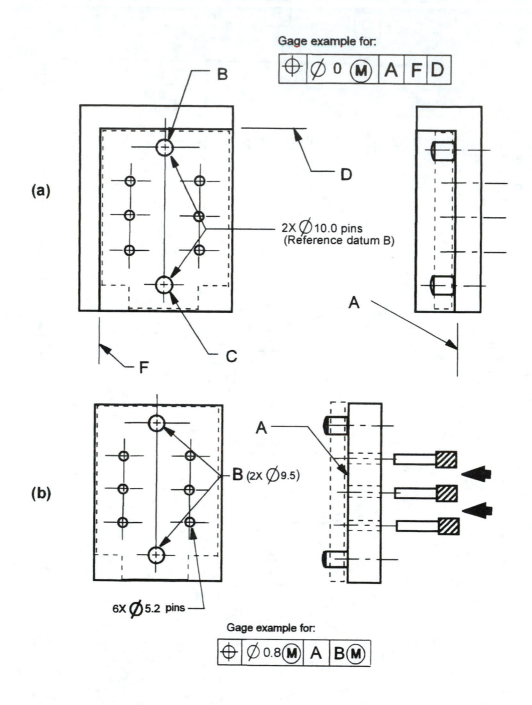

Gage example for:

⊕ | ⌀ 0 Ⓜ | A | F | D

(a)

B

D

2X ⌀ 10.0 pins
(Reference datum B)

A

F

C

(b)

A

B (2X ⌀ 9.5)

6X ⌀ 5.2 pins

Gage example for:

⊕ | ⌀ 0.8 Ⓜ | A | B Ⓜ

Figure 3-28 (continued) Gage explanations.

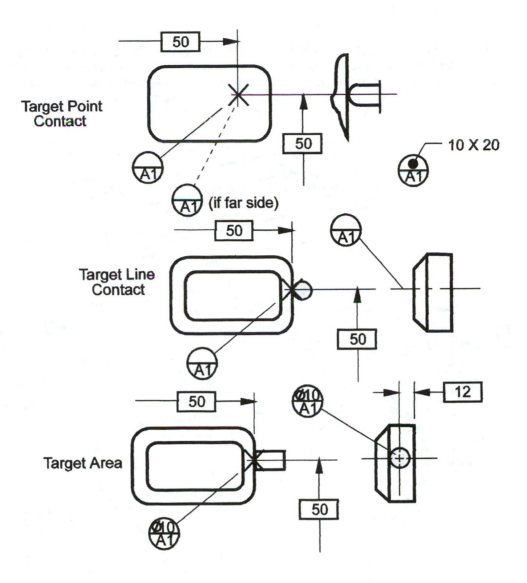

Figure 3-29 Datum targets.

TARGET POINTS AND SIMULATED DATUMS

Figure 3-30 illustrates a part which is restrained in a fixture by use of vee locators and target points. This practice will equalize casting, forging, or molding errors so that holes or other features may be centered in the rough part. The plane that is created by this method more appropriately exists in the fixture rather than on the part. The term for the plane is *simulated datum*. Simulated datums are appropriate when it is desired to equalize production errors or when feature datums are unavailable. Figure 3-30 shows the part secured in the fixture with production errors distributed about centerplane B.

Note the datum target leader lines in the plan view of Figure 3-30. Because the target point is on the far side of the part, the line is shown as a hidden line. If vee locators were fixed at the left side and were adjustable on the right, the tertiary datum could be established on the left side of the part as shown. If both right and left sides of the fixture were adjustable, as shown in Figure 3-31, a simulated datum could exist at the centerplane of the fixture as shown.

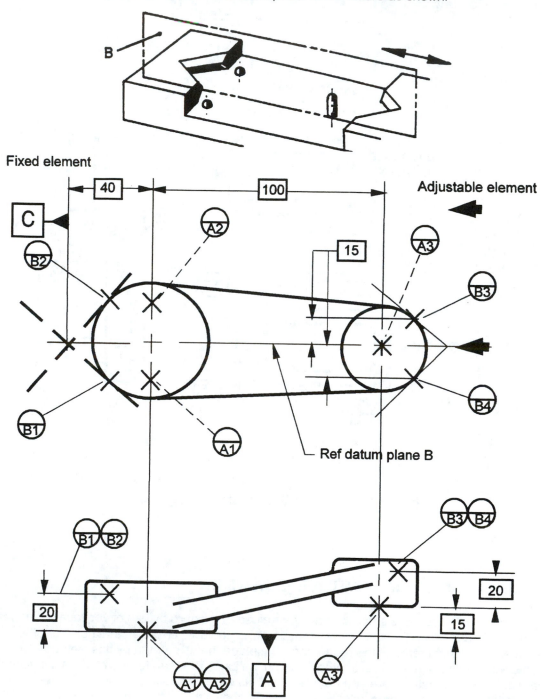

Figure 3-30 Simulated datum plane: Inspection or processing equipment.

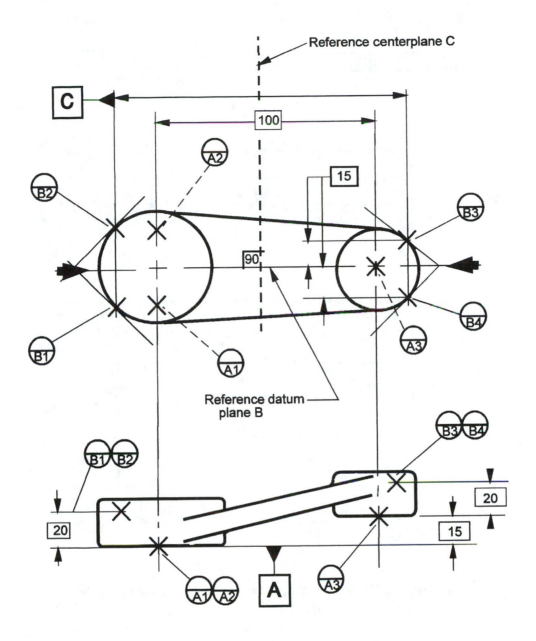

Figure 3-31 Simulated datum - two adjustable elements.

PRIMARY DATUM AXIS

Recall the discussion on datum axes and target points. Figure 3-32 illustrates the use of target points to create the primary datum axis. From the diameter length to surface ratio, we see that the greater design influence appears to rest with the diameter. Due to the length of primary datum feature A, it may be impractical (or impossible) to locate on the entire diameter for evaluation or measurement. In these cases, *target points* at each end of diameter A could aid in establishing axis A (see Figure 3-32a). In this figure, a partial datum has been created, using only points (partial) on the datum feature. In Figure 3-32b, the same technique is used; however, because datum features A and B are of different sizes (two separate features), the targets establish datum axis A-B. The shaft end in both instances is the secondary datum.

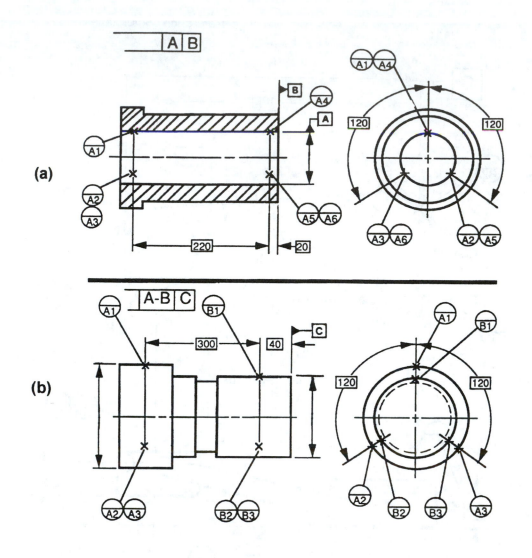

Figure 3-32 Primary datum axis: target points.

EQUALIZING ERROR AND DATUM TARGETS

Sometimes the very nature of a product makes producing a quality part very difficult. An example might be a stamping or molding/casting where production errors tend to accumulate. In some cases, industry standards refer to tolerances *per unit length,* allowing errors to compound or grow, according to length. One method of coping with this problem is to simulate datums using fixture centerplanes to equalize error. The example in Figure 3-33 shows a long cover, possibly a casting, with datum targets in both X and Y directions. These targets may simulate hydraulic or mechanical locators. They could actuate in the fixture, positioning the part so that when dimensioned and located from the centerplanes, production errors are halved. The datum framework is mutually 90 degrees (by definition) and is controlled in the fixture. Another possible method is to used a datum dimensioning technique to control symmetry, similar to that shown in Figure 7-13.

104

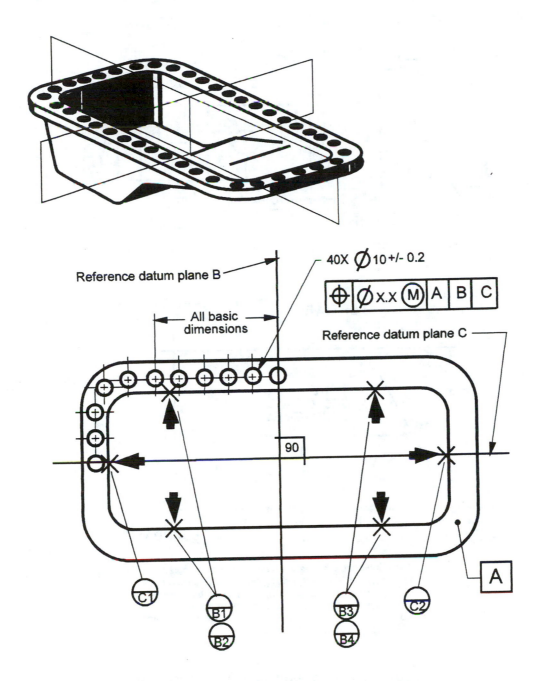

Figure 3-33 Datum targets and datum simulators.

INCLINED DATUMS

Figure 3-34 illustrates a condition where the tertiary datum is established as mutually perpendicular to the other two datums, but due to the shape of the part, it is necessary to rotate it about the intersecting planes to locate the true geometric counterpart of surface C. The fixture would locate the part on primary datum surface A and secondary datum surface B, with the part sliding to be locked in on angular tertiary datum surface C.

105

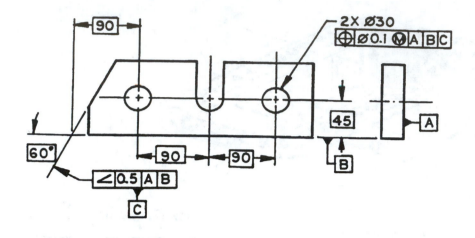

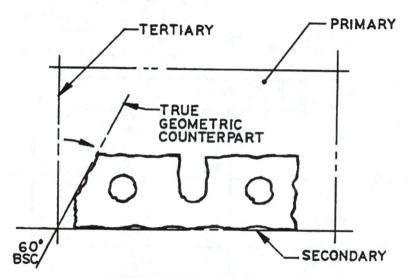

The contacting plane is oriented at the basic angle of the feature. The corresponding plane of the datum reference frame is rotated through this same basic angle to be mutually perpendicular to the other two planes. For this method of establishing a datum reference frame. The angle is indicated as basic.

Figure 3-34 Inclined datum frameworks.

MULTIPLE DATUM FRAMEWORKS

We will undoubtedly encounter designs with requirements relating to more than one set of datum framework. The requirements need to be separated and the design thought through, from the most to the least critical, in order of design influence. Each framework (differing order of datums) constitutes a separate functional and gaging requirement. One gage or measurement setup will not satisfy all the requirements of Figure 3-35.

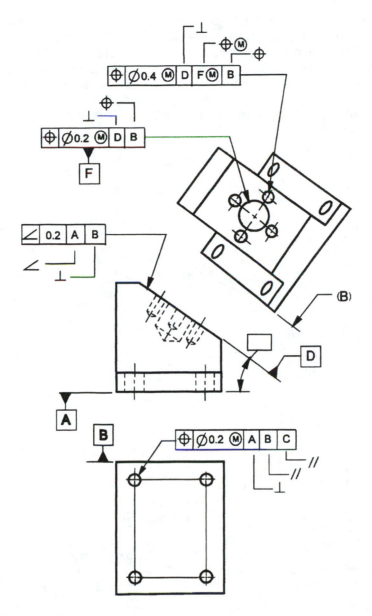

Figure 3-35 Multiple datum frameworks.

MATHEMATICALLY DEFINED AND COMPLEX SURFACES

In many cases, parts will not have features with clearly defined flat or cylindrical surfaces. Airfoils and turbine blades are good examples. Still, a method of defining a datum framework is required to communicate design requirements and for measurement. The mathematically defined surface illustrated in Figure 3-36 is an example of a warp surface located in a datum framework. Points on the surface, located with basic dimensions, could then be related to the framework. Profile tolerances could then be applied to the surface. other techniques are possible. See also Figure 5-10.

Figure 3-37 illustrates a mathematically defined curved primary datum surface. Three holes are located with the hole centers normal to the datum surface. In this figure, the basic dimensions between holes, and to other features, are designated as arc distances along the datum surface. The three holes as a pattern are datum features B, and, together with the primary datum surface, they constitute datum framework A, B. Because datum all features are specified MMC, it is possible to devise a functional gage design with MMC pins for datum and feature holes. The hole depth in the gage would be equal to the feature or flange thickness, and all gage pins would be loose pins.

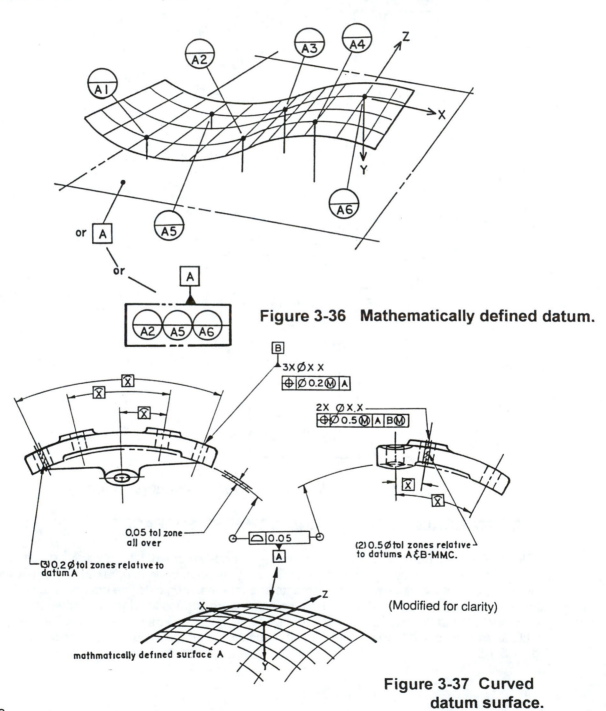

Figure 3-36 Mathematically defined datum.

(Modified for clarity)

Figure 3-37 Curved datum surface.

DATUMS SUMMARY

Datum features shall be accurate, accessible, and of reasonable size to ensure repeatable measurement results.

A datum feature may be a single feature or a pattern of features.

Secondary and tertiary datums, when specified, must be controlled to the primary datum.

Both datum features as well as datum targets may be referenced in the feature control frame.

Datum target points need not lie in the same plane.

Tooling tolerance is implied for datum target point, line, or area location.

The ISO datum feature symbols are now employed in ASME Y14.5M.

Secondary and/or tertiary datum features of size must be simulated at their *virtual condition*.

Surfaces defined by mathematical data can be designated as datum features.

Equalized centerplanes and axes may be specified as datum centerplanes and axes under certain conditions.

EXERCISE 3   Datums

1. In order of importance, datums are _____, _____, and _____.

2. Occasionally, datum features are not possible or practical to use. In this case, _____, _____, or _____ may be substituted for datum features.

3. Ideally, datums should be determined and specified based on _____.

4. The figure below represents a _____ datum.

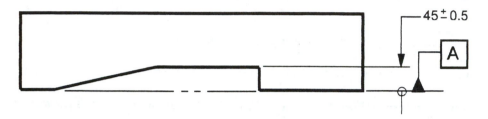

5. The datum that stops rotation of a cylindrical part in a setup is normally the _____ datum.

6. The use of opposite ends of a shaft simultaneously for datum purposes is shown by this callout:

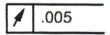

7. The datum framework consists of ____ mutually _____ _____.

8. In an MMC functional receiver gage, secondary and tertiary datum features of size must be applied at their _____.

9. In the figure below, specify the top mounting surface, in the right side view, as datum A. Specify the flat surface at the bottom of the part as datum B and the 20 mm diameter pilot as datum C.

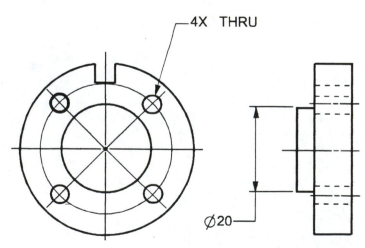

EXERCISE 3 Datums (cont)

In the previous figure, specify the 20 diameter pilot as datum C. Specify the slot centerplane as datum D.

Which datum features are subject to size variation and virtual conditions _____?

Relative to the four small holes, from datums A, C, and D, what sequence and modifier symbols appear the most logical if functional gaging is to be used?

4X THRU

⊕	⌀ x.x Ⓜ	

True or False:

1. Datum accuracy and accessibility are equally important. T   F
2. Datum order is determined by functional influence. T   F
3. The datum reference frame provides stability for measurement. T   F
4. Datum planes have length, width, and depth. T   F
5. Actual part surfaces are called datum features. T   F
6. Datum symbols should be shown to centerlines and/or centerplanes. T   F

The part shown below operates on a shaft, with adjacent levers, to actuate movement of intake and exhaust valves. In your best judgment, what should be datum A? What then would be secondary and tertiary datums B and C? How would you complete the control callous for the oil and adjusting screw holes, if functional pin gaging is to be used? Show the secondary datum to be perpendicular to the primary datum within 0.5 RFS.

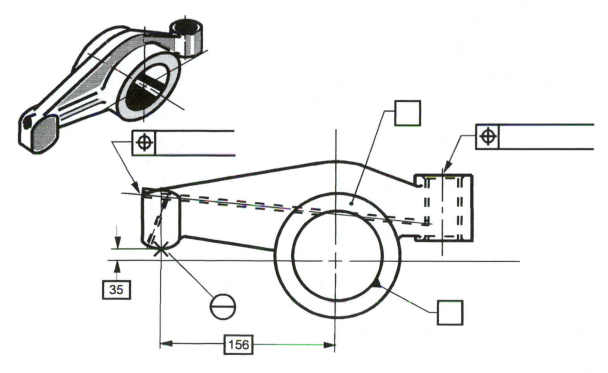

35

156

4 ORIENTATION CONTROLS

PERPENDICULARITY   ANGULARITY   PARALLELISM

⊥   ∠   //

Orientation controls are datum related. *Perpendicularity, angularity, and parallelism* (see Figure 4-1) are in the category of *orientation controls.* These controls may be applied with material modifiers MMC or LMC; however, RFS is implied unless otherwise stated (ref. Rule 2). If applied to a planar surface, orientation controls may also control *form,* such as *straightness* or *flatness*, within the orientation control tolerance value. The orientation tolerance limits must be contained within the size tolerance limits, per Rule 1. If applied to a feature of size, with an axis or centerplane, a *virtual condition* may allow the feature inner/outer boundary limit to extend beyond the size limits, and therefore Rule 1 could not apply.

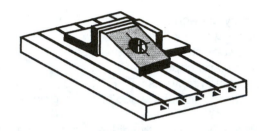

Figure 4-1 Features related to datum features.

The design application should be carefully studied before applying a control to a feature centerplane or axis, because there may be more than one feature sharing the same axis or centerplane. It may not be clear for which feature the control was intended, that is, a feature with a counterbore or holes in line hidden behind other feature holes.

PERPENDICULARITY

Perpendicularity is the condition of a surface, centerplane, or axis at a right angle (90 degrees) to a datum plane or axis. A perpendicularity tolerance control specifies one of the following:

1. A tolerance zone defined by two parallel planes perpendicular to a datum plane or axis within which a surface or median plane of the considered feature must lie.
2. A tolerance zone defined by two parallel planes perpendicular to a datum axis within which the axis of the considered feature must lie.
3. A cylindrical tolerance zone perpendicular to a datum plane within which the axis of the considered feature must lie.
4. A tolerance zone defined by two parallel lines perpendicular to a datum plane or axis, within which an element of the surface must lie.

Figure 4-2 illustrates surface perpendicularity to only one datum reference allowing the feature surface to be skew, within size limits, to other surfaces. This condition represents limited *degrees of freedom*. With only one datum invoked, there exists full freedom of disorientation to other datum surfaces, which may require a secondary datum or other orientation control.

112

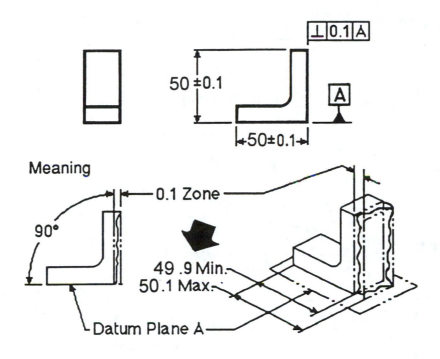

Figure 4-2 Perpendicularity of a surface.

Figure 4-3 illustrates perpendicularity of an axis at both RFS and MMC. Note that with RFS specified, the tolerance is always the same (.003), whereas the *Outer Boundary* is variable (.503-.501). With MMC, however, the tolerance is variable depending on size, whereas the *virtual mating size* is constant at .503, the worst-case (closest) fit. In the MMC example, perpendicularity is applied to a feature of size; therefore, *a virtual condition* may occur.

Figure 4-4a shows the use of zero tolerance at MMC and the effect of a restricted tolerance control, *which does not allow the full bonus tolerance to be applied.* Figure 4-4b illustrates the application and control of radial line elements. If the qualifying note EACH RADIAL ELEMENT were not used beneath the control frame, the callout would apply to the entire surface.

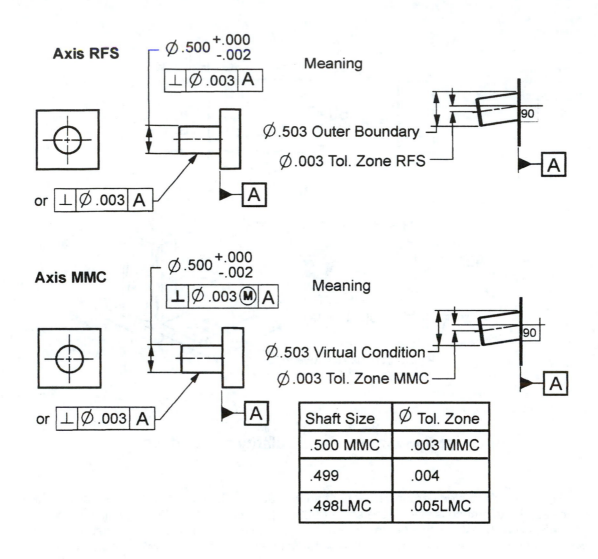

Figure 4-3 Perpendicularity: axis RFS and MMC to datum surface.

114

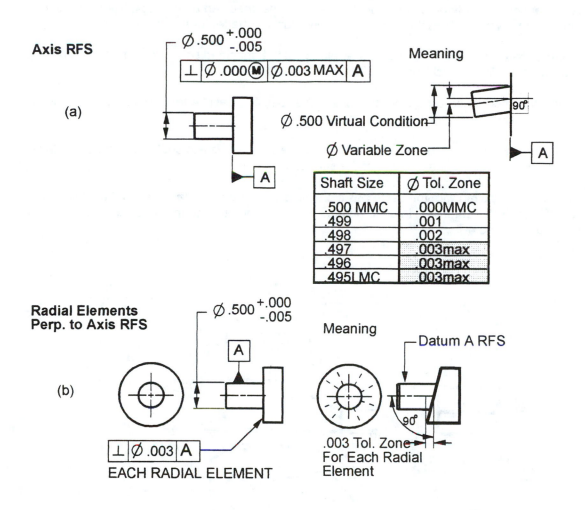

Axis RFS

$\emptyset.500^{+.000}_{-.005}$

| ⊥ | $\emptyset$ .000 Ⓜ | $\emptyset$ .003 MAX | A |

(a)

Meaning

$\emptyset$ .500 Virtual Condition

$\emptyset$ Variable Zone

90°

A

A

Shaft Size	$\emptyset$ Tol. Zone
.500 MMC	.000MMC
.499	.001
.498	.002
.497	.003max
.496	.003max
.495 LMC	.003max

Radial Elements Perp. to Axis RFS

$\emptyset.500^{+.000}_{-.005}$

A

(b)

| ⊥ | $\emptyset$ .003 | A |

EACH RADIAL ELEMENT

Meaning

Datum A RFS

90°

.003 Tol. Zone
For Each Radial
Element

Figure 4-4 Perpendicularity: axis MMC and restricted tolerance. Radial elements to datum axis.

115

ANGULARITY

Angularity is the condition of a surface, centerplane, or axis at a specified angle (other than 90 degrees) from a datum plane or axis. An angularity tolerance control specifies one of the following:

1. A tolerance zone defined by two parallel planes at the specified basic angle from the datum plane or axis within which the surface or centerplane of the feature must lie.

2. A tolerance zone defined by two parallel planes at the specified basic angle from a datum plane or axis within which the axis of the feature must lie.

3. A cylindrical tolerance zone whose axis is at the specified basic angle from a datum plane or axis within which the axis of the feature must lie.

4. A tolerance zone define by two parallel lines at the specified basic angle from a datum plane or axis within which a line element of the surface must lie.

Angularity is applied with the same basic principles as perpendicularity, except angularity is intended for use with features related to datums at orientation other than 90 degrees. See Figures 4-5, 4-6, and 4-7.

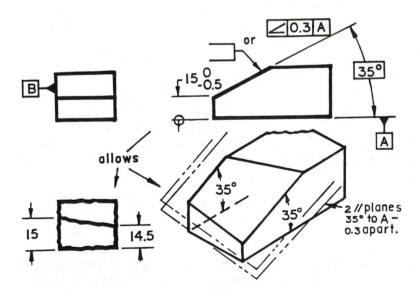

Figure 4-5 Angularity of a surface.

116

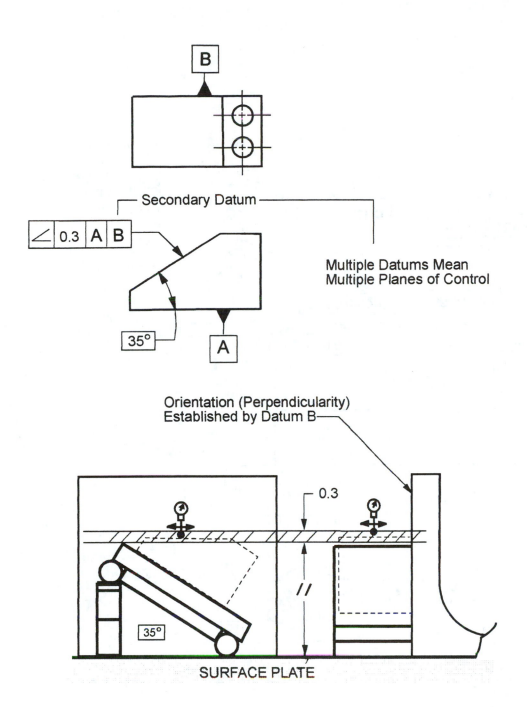

Figure 4-6   Angularity with secondary
datum control specified.

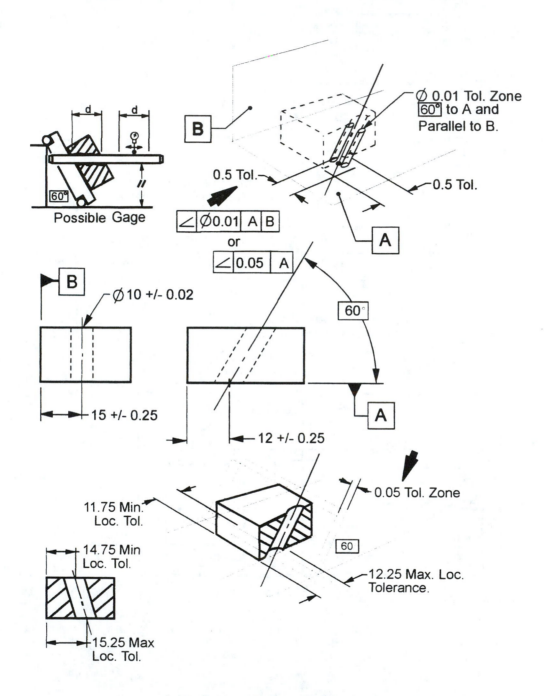

Possible Gage

Ø 0.01 Tol. Zone
60° to A and
Parallel to B.

B

0.5 Tol.

0.5 Tol.

∠ | Ø0.01 | A | B

or

∠ | 0.05 | A

A

60°

A

B

Ø 10 +/- 0.02

15 +/- 0.25

12 +/- 0.25

0.05 Tol. Zone

60

11.75 Min.
Loc. Tol.

14.75 Min
Loc. Tol.

12.25 Max. Loc.
Tolerance.

15.25 Max
Loc. Tol.

Figure 4-7 Angularity of a feature axis.

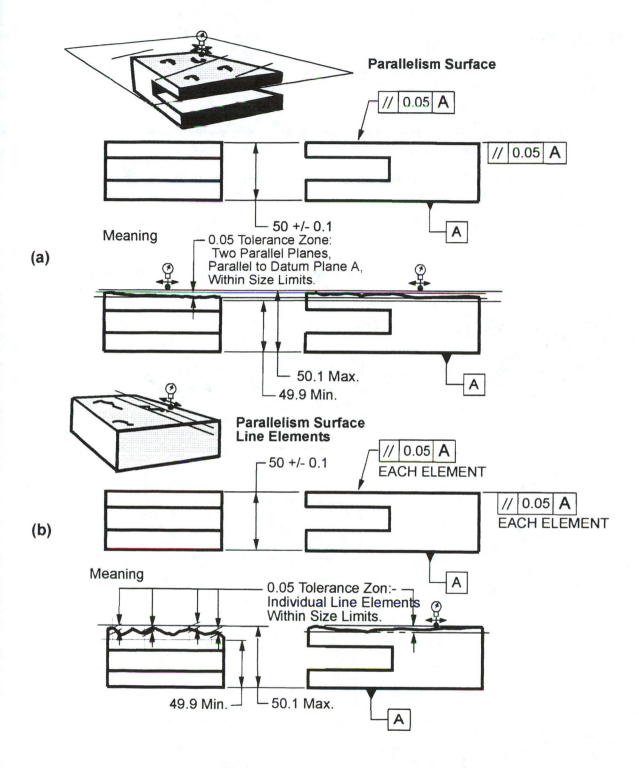

Parallelism Surface

// | 0.05 | A

// | 0.05 | A

50 +/- 0.1

A

Meaning

(a)

0.05 Tolerance Zone:
Two Parallel Planes,
Parallel to Datum Plane A,
Within Size Limits.

50.1 Max.
49.9 Min.

A

**Parallelism Surface
Line Elements**

50 +/- 0.1

// | 0.05 | A
EACH ELEMENT

// | 0.05 | A
EACH ELEMENT

(b)

A

Meaning

0.05 Tolerance Zon:-
Individual Line Elements
Within Size Limits.

49.9 Min. ── 50.1 Max.

A

**Figure 4-8   Parallelism surface and
surface elements.**

PARALLELISM

Parallelism is the condition of a surface or centerplane equidistant at all points from a datum plane or an axis equidistant along its length to a datum plane or axis. A parallelism tolerance control specifies:

1. A tolerance zone defined by two parallel planes parallel to a datum plane or axis within which the surface or centerplane of the feature must lie. See Figure 4-8a.

2. A tolerance zone defined by two parallel planes parallel to a datum plane or axis within which the axis of the feature must lie.

3. A cylindrical tolerance zone parallel to a datum plane or axis within which the axis of the feature must lie. See Figure 4-9.

4. A tolerance zone defined by two parallel lines parallel to a datum plane or axis within which an element of the feature surface must lie. See Figure 4-8b.

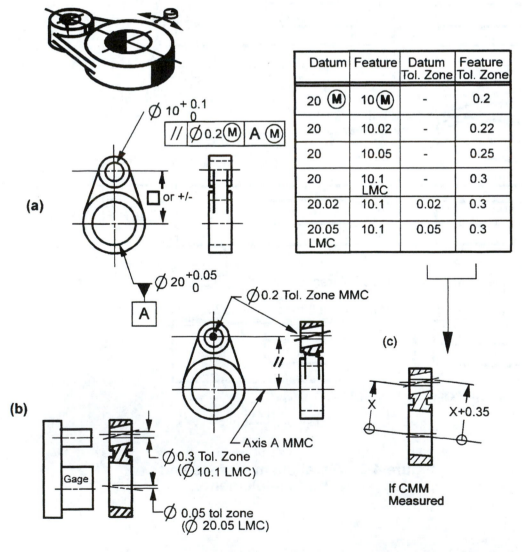

Datum	Feature	Datum Tol. Zone	Feature Tol. Zone
20 Ⓜ	10 Ⓜ	-	0.2
20	10.02	-	0.22
20	10.05	-	0.25
20	10.1 LMC	-	0.3
20.02	10.1	0.02	0.3
20.05 LMC	10.1	0.05	0.3

120   **Figure 4-9 Parallelism: axis to axis MMC/RFS (MMC shown).**

Observe that *parallelism, perpendicularity, and angularity* have similar principles governing their use. The placement location of the feature control frame is optional. When applied to a feature surface, the controlled feature must be within size limits. Also notice the impact of MMC applied to a datum feature of size, as shown in Figure 4-9a. A bonus tolerance is available as the datum feature departs from MMC. This tolerance is automatically accommodated by a functional gage as shown in Figure 4-10, but must be allowed for if measured with a CMM, or other open setup inspection process. See Figure 4-9c.

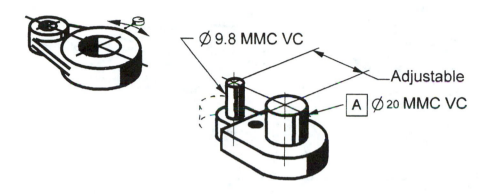

Figure 4-10 Possible functional gage for Figure 4-9b.

Tangent Plane Concept

The *tangent plane* concept was added to the ASME Y14.5-1994 standard. It differs from parallelism in that the *tangent plane control* applies to a plane contacting the high points of a surface and requires the plane be within the parallelism limits; surface elements, however, are allowed to deviate as long as they remain within the size limits. Parallelism requires *all surface elements* to be within the parallelism limits as well as within size limits. This subtle design difference should be reviewed carefully. This concept was felt necessary for use on extremely long surfaces, or surfaces with great surface area, where the contacting planes were critical but localized surface form deviations were not. See Figure 4-11.

When it is necessary to control the orientation of a feature surface as established by the high, or contacting, points of that surface, the *tangent plane* symbol is added within the feature control frame. In Figure 4-11, a plane of contact shall lie within two parallel planes 0.1 apart, parallel to datum surface A. All elements of the surface must be within the size limits.

ORIENTATION CONTROLS SUMMARY

Orientation controls may be applied to feature surfaces or axes/centerplanes, relative to a datum reference. When applied to a size feature axis or centerplane, violation of the MMC envelope may occur. Carefully consider the design before applying a control to an axis, centerplane, or plane because there may be more than one feature sharing the axis or plane. The drawing should be clear as to which feature the control is intended, such as multiple interrupted surfaces in alignment, multiple aligned holes hidden from view, and counterbores or counterdrilled holes features. The *tangent plane* concept differs from parallelism, in that the surface form is *not* controlled.

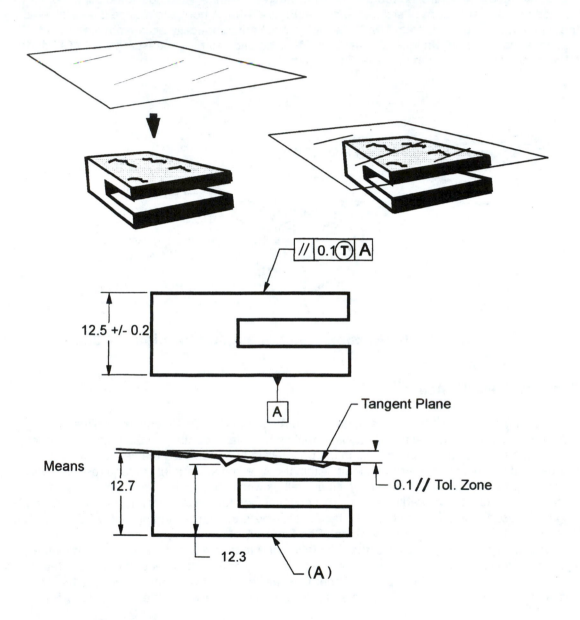

Figure 4-11   Tangent plane concept.

EXERCISE 4-1 Parallelism

True or False, or fill in.

1. Parallelism is applied MMC unless otherwise specified. T F

2. Parallelism is to be related to a datum. T F

3. Relative to a flat surface, Parallelism will also control _____ and _____.

4. Parallelism is to be contained within size tolerance. T F

5. When applied to a flat surface, the parallelism tolerance may be additive to size tolerances. T F

6. A parallelism tolerance control specifies a +/ - tolerance zone. T F

7. In the figure below, illustrate a parallelism tolerance control for the top surface of 0.2 to datum A. <u>Draw the tolerance zone</u>.

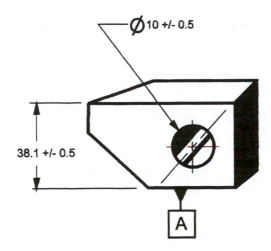

8. Illustrate the axis of the hole to be parallel to datum A within a 0.1 diameter tolerance zone at MMC.

9. What is the hole tolerance zone size, if the hole diameter is at LMC?

10. What is the maximum permissible value for the 38.1 mm dimension?

11. What is the hole MMC virtual condition limit?

EXERCISE 4-2 Perpendicularity

True or False, or fill in.

1. Perpendicularity is applied MMC unless otherwise specified. T   F

2. Perpendicularity is independent from datum relationships. T   F

3. When applied to a feature _____ or _____, a virtual condition may be invoked.

4. When applied to a flat surface, perpendicularity will also control _____ and _____ within the limits of the perpendicularity control.

5. Perpendicularity may be applied to cylindrical features and their relationships.   T   F

6. The figure shown below needs a 90 degree dimension.   T   F

7. In the figure below, illustrate surface X to be perpendicular to datum surface A within 0.1. <u>Draw the tolerance zone.</u>

8. Indicate the hole to be perpendicular to datum surface A within a diameter tolerance zone of 0.2 at MMC.

9 What is the tolerance zone if the hole diameter is 10.5?

10. What is the hole MMC Virtual Condition limit?

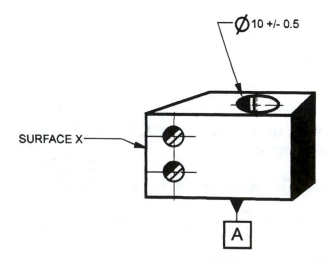

EXERCISE 4-3 Angularity

True or False, or fill in.

1. Angularity is applied RFS unless otherwise specified. T   F

2. Angularity is not related to datums. T   F

3. Relative to a flat surface, an angularity tolerance zone is the space between two parallel planes. T   F

4. Angularity may be applied to a flat surface at MMC. T   F

5. Angularity is not appropriate for 90 degree relationships. T   F

6. To control the hole, a feature of size, in the figure below, the use of MMC will allow a bonus tolerance. T   F

7. For the figure below, illustrate the hole to have a 60 degree basic relationship to datum A, and indicate an angularity tolerance control of 0.1 diameter to datum surface A. <u>Draw the tolerance zone.</u>

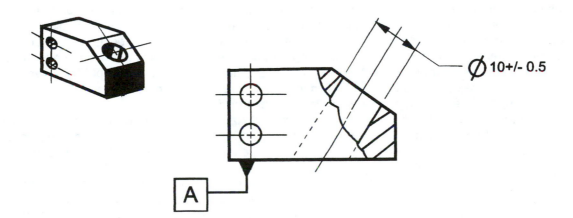

8. The LMC hole size is _____.

9. What is the MMC virtual condition hole size?

10. Illustrate the inclined surface to be 25 degrees basic to surface A, with an angularity tolerance control of 0.2 to datum A.

5 PROFILE CONTROLS

Profile tolerancing is used to specify a permissible deviation from a desired *profile*, usually an irregular shape, where other controls are inappropriate. See Figure 5-1. The *profile* tolerance specifies a uniform boundary along or about the true profile within which the surface, or elements of the surface, must lie. See Figures 5-2 through 5-5. It is used to control *form* or combinations of *size, form, orientation* and *location*. Where used as a refinement of size, the profile tolerance must be contained within the size limits.

Profile - Line   Profile - Surface

Figure 5-1   Profile control symbols.

Profile tolerancing is considered one of the most versatile controls available and may be used with or without datums, to control size, to control curved or warped surfaces, or to control the coplanar relationships of flat surfaces.

In addition, when used as a composite control, profile contains a degree of *position* and *form* control within the same callout. Profile is generally applied RFS to features, but may be applied RFS or MMC to the datum references.

EXAMPLES OF PROFILE CONTROLS

For *profile of a line*, the tolerance zone exists at any section parallel to the view in which it is presented. For *profile of a surface*, the tolerance zone applies to the entire surface in the view in which it is presented. The examples in this chapter use surface control symbols, but line profile symbols could also have been used. Normally the profile tolerance zone is equally disposed about the true basic profile and is identified with basic dimensions. To dispose the tolerance unequally, it is required to illustrate pictorially and/or dimensionally how the tolerance is to be applied, as shown in Figure 5-2. Note also that this method may be used to apply all tolerance *outside or inside* the true profile, as shown in Figures 5-2b and c. Unless specifically stated otherwise, the interpretation in Figure 5-2a applies to profile controls.

Figure 5-3 illustrates the use of profile to control size also. No datums are invoked. With the use of the *all-around symbol*, however, the profile tolerance applies all around the true profile shape, in the view presented. Because no datums are specified, *no automatic controls* to other surfaces (perpendicularity) exists.

Figure 5-4 is similar to Figure 5-3 except that it is modified to illustrate the use of a datum framework. Two profile controls are specified: the curved ramp surface between points X and Y and the vertical surface between points Y and Z. In this illustration, profile is used to control *form* and *orientation (perpendicularity)*. Because datums have been invoked, the profile tolerance zones must be oriented to the primary datum B, to control perpendicularity, and located from secondary and tertiary datums A and C.

Figure 5-5 illustrates the use of profile as a refinement of size, with the profile tolerance zone free to float within the size tolerance zone. Without datums specified in the control frame, the profile tolerance may twist or tilt, within the size limits. Additionally, the designer has specified the line elements of the curved surface between points X and Y to be parallel to datum surface A. Combinations of multiple controls such as this are often used to achieve multi-directional control. In the example shown, the 35 +/- 0.5 dimension not only establishes the size limits but also establishes the origin of the R12 basic profile. '

126

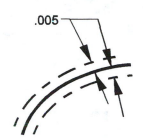

⌒ **Line Element**
⌓ **Surface**
Always RFS

Profile tolerancing specifies a uniform boundary along or about the true profile within which the elements of the feature surface must lie. Profile may also control form, combinations of size, form orientation and location, with or without datum references. Where used as a refinement of size, the tolerance must be contained within the size limits

.005

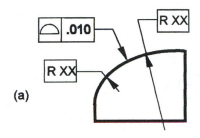

(a)

Means

An equally disposed or bilateral tolerance zone about the true profile.

or if so specified

.007
.003

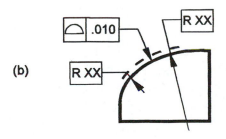

(b)

Means
An unilateral, tolerance zone "outside" the basic profile.

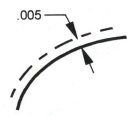

.005

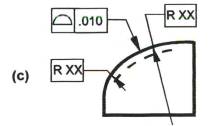

(c)

Means

An unilateral tolerance zone "inside" the basic profile.

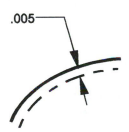

.005

FIG 5-2 Profile Controls

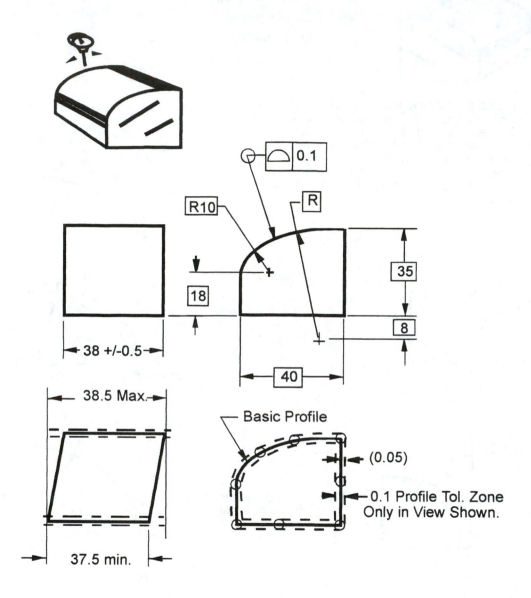

Figure 5-3   Profile all-around surface (also controls size).

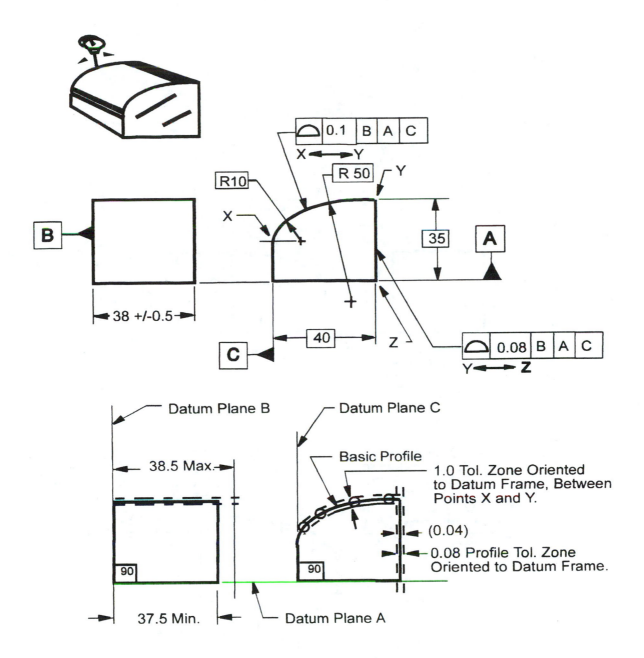

**Figure 5-4   Profile control to datum frame
(also controls form and orientation).**

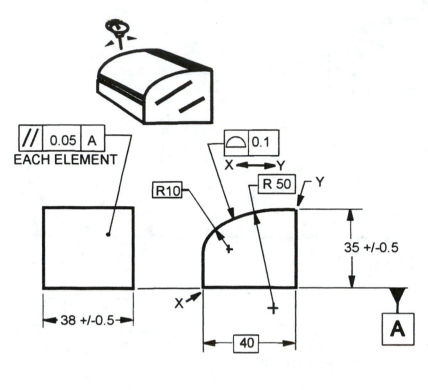

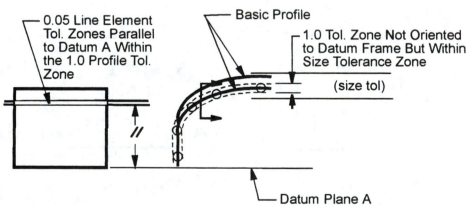

Figure 5-5 Profile tolerance combined with size and other orientation controls.

PROFILE APPLIED: SECONDARY AND TERTIARY DATUMS

Figure 5-6a illustrates the use of two holes, of a pattern of four, as secondary and tertiary datums B and C. The internal and external boundaries are controlled to the datum frame by use of profile at MMC. The datum holes are controlled at MMC; therefore, a functional fixed pin gage may be used as shown. In Figure 5-6b, the profile control is specified RFS, thus requiring an adjustable fixture and/or gage pins. Effects of RFS and MMC on datum features of size were covered in the datum section. Figure 5-6 might be typical of some gasket or shim stock application dimensioning.

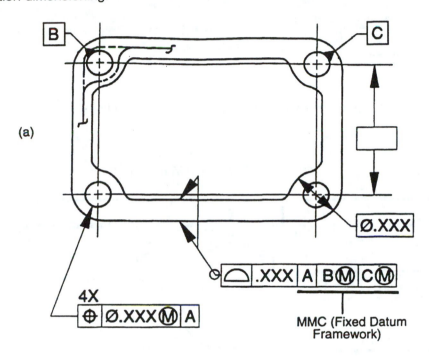

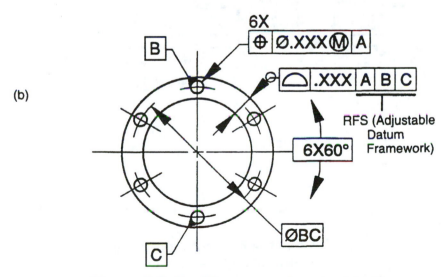

Figure 5-6 Profile: secondary and tertiary datums RFS versus MMC.

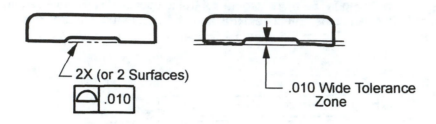

Figure 5-7 Profile control: coplanar surfaces.

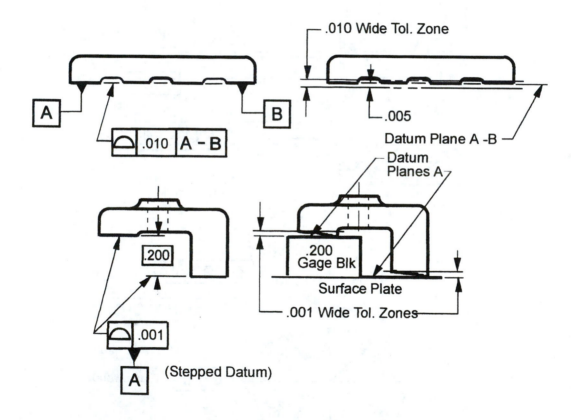

Figure 5-8 Profile controls: coplanar and stepped surfaces.

As previously discussed, profile may be used to control coplanar surfaces. Coplanarity is the condition of two or more surfaces having all elements in one plane. Profile tolerance control may be used when it is desired to control multiple interrupted surfaces as a single continuous surface. In this case, a profile control applied to multiple surfaces achieves the same result as flatness control applied to a single surface. The tolerance zone is defined as two parallel planes within which all surface elements of all surfaces must lie. A common technique on older drawings was to specify a note such as, " Surfaces A and B to lie in the same plane within .010." The use of the profile control as shown in Figure 5-7 simplifies and standardizes this requirement.

When establishing a datum from an interrupted surface, the datum is considered as continuous, unless otherwise noted. Note that *flatness* applies only to single surfaces.

The profile control may also be used for multiple, coplanar surface controls. Figure 5-8 illustrates the use of any two coplanar feet selected as datums A and B, with the remaining feet controlled relative to these datums. As shown in Figure 5-8, profile may also be used to control parallel, offset feet. The figure has created a *stepped datum* condition, with the .001 tolerance applying to both feet, considered as one. Placement of the datum symbol further confirms this.

PROFILE APPLICATION

Figure 5-9 illustrates profile of a surface applied to a cylindrical surface. As shown, the tolerance zone exists at the feature surface, for full feature length, and because a datum is referenced, it is relative to the datum axis A. In the example shown, similar results could be obtained by use of *total runout*. Note, however, the additional flexibility of the profile control to also control form or orientation.

Figure 5-10 illustrates the use of profile to define and control a curved surface (taken from the datum section), which will be used as a datum surface, once verified.

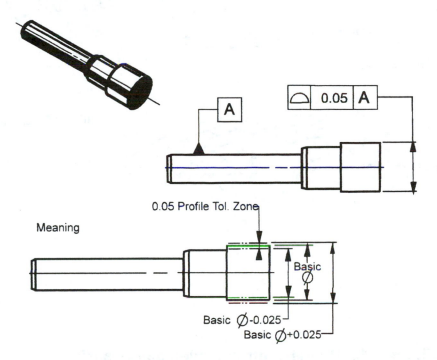

As illustrated, profile callout controls size, form (cylindricity), orientation (parallel to axis A), and location (from axis A).

Figure 5-9   Profile applied to cylindrical feature.

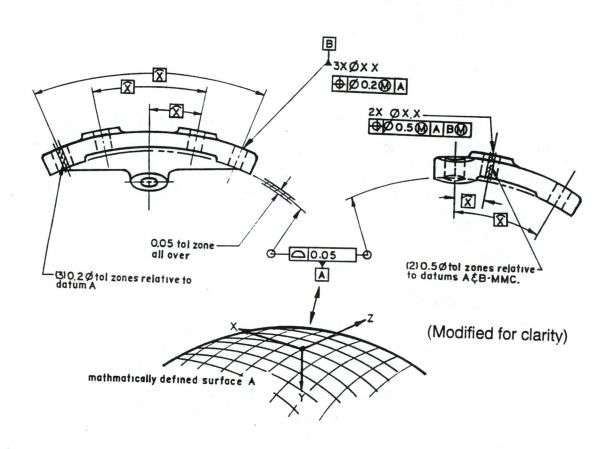

B

3X Ø X X

⊕ Ø 0.2 Ⓜ A

2X Ø X.X

⊕ Ø 0.5 Ⓜ A B Ⓜ

0.05 tol zone
all over

(3) 0.2 Ø tol zones relative to
datum A

(2) 0.5 Ø tol zones relative
to datums A & B-MMC.

⌓ 0.05

A

(Modified for clarity)

Z

X

Y

mathmatically defined surface A

Figure 5-10 Profile control: curved datum surface.

COMPOSITE PROFILE CONTROLS

Composite Profile tolerancing is much like composite positional tolerancing. The upper control callout establishes the *location boundary* and governs the *location* of the profile shape. The lower control callout governs the refinements of size, form, and orientation and specifies a smaller tolerance zone that must fall within the upper location boundary. The actual feature surface must therefore lie within both zones.

As stated, composite profile tolerancing is much like composite position tolerancing except that with profile, the feature has a *boundary tolerance zone,* whereas with position, we have an axial or centerplane tolerance zone. Figure 5-11 shows a *composite profile control.*

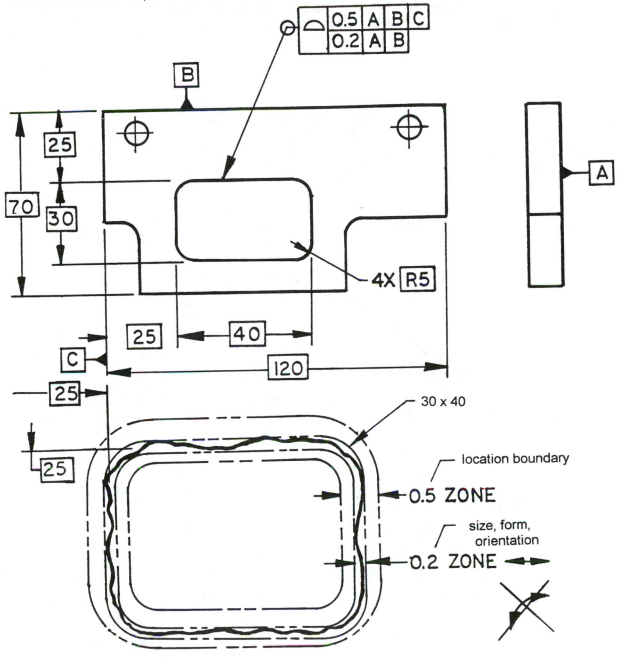

Figure 5-11 Composite profile control.

135

The location tolerance of 0.5 mm is relative to the datum framework A, B, C and governs the location of the controlled shape. The lower 0.2 control governs the size, form, and orientation, tolerance limits that must fall within the upper limit of 0.5. Datums A and B are invoked in the lower control, which means that the lower control tolerance zone may slide from side to side and up and down, but must remain perpendicular to A and parallel to B. The 0.2 tolerance zone must remain within the 0.5 zone.

MULTIPLE PROFILE CONTROLS

Adapting the the datum and profile control principles established earlier, it is possible to use profile for various design requirements. Figure 5-12 illustrates the use of multiple profile controls on a complex shape, defined by mathematical means.

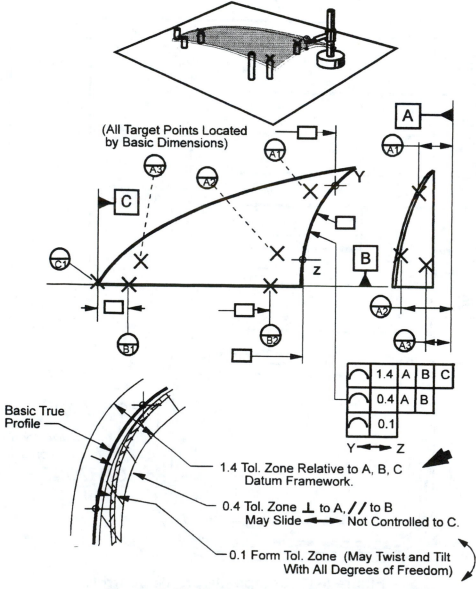

Figure 5-12 Multiple controls: complex feature.

Composite controls have been extended to include orientation (datum reference) and form (no datum reference). The upper control establishes the datum framework and therefore controls location. Subsequent controls establish orientation and/or form controls, depending on datum references applied.

The second tolerance control (0.4) may slide as shown in Figure 5-12, parallel to datum B, and remain perpendicular to A, within the limits established by the 1.4 location zone. The lower control is a further refinement of form, but with no datum reference. The 0.1 tolerance zone is free to twist and tilt, within the limits established by the 0.4 zone. The 0.4 zone has full degrees of freedom.

BOUNDARY CONCEPT

When it is desired to control the size and form, as well as location of irregular shapes to a datum framework, profile tolerancing in combination with position tolerancing may be used. Position tolerancing is generally applied to features that have an axis or centerplane, such as holes, slots, and keyways. If the feature is irregular and does not fit the definition of a *feature of size*, cylinder, sphere, or two opposed feature surfaces, consideration should be given to the use of profile tolerancing to control shape and size, along with position tolerancing to control location. This concept is illustrated in Figure 5-13.

The *boundary concept* is used when tolerance zones are to be verified by measurement of MMC virtual condition boundaries generated. This concept was originally noted in ANSI Y14.5 for slots and shafts, and has been expanded to include other shapes not always considered features of size in the past. The concept is the same as for a cylindrical feature oriented or positioned to a datum framework. If a *virtual condition boundary* can be calculated for the controlled feature, a gage can be generated for that feature boundary as well. In such cases where the boundary is to be measured or gaged (in lieu of a tolerance zone at the axis or centerplane), the word "BOUNDARY" is placed beneath the feature control frame.

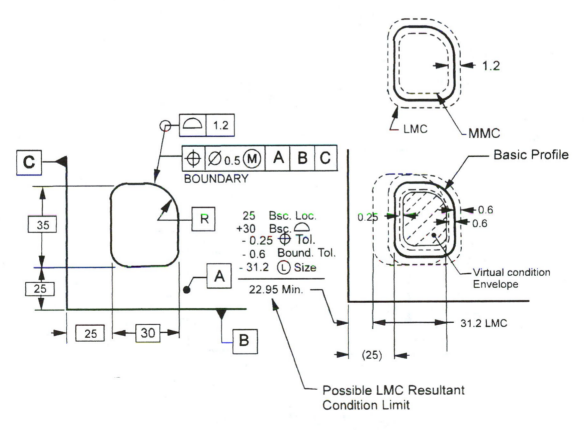

Figure 5-13 Boundary Concept

137

As shown in Figure 5-13, the profile tolerance controls the shape and size and is not related to the datum framework, other than perpendicularity to A. The position control locates the shape at the basic dimensions shown. The MMC principle allows a bonus tolerance when the feature departs from MMC and gets larger. As the feature reaches LMC condition, the resultant condition creates a minimum wall thickness, which should be considered in the design. One wall thickness result is 22.95, as shown in Figure 5-13.

Figure 5-14 offers further explanation of the *boundary principle* when used with other combinations of profile, plus-minus or position controls, to both feature surfaces and axes. Figure 5-14a illustrates plus-minus size dimensioning along with positional tolerance controls to a feature axis. This dimensioning/tolerancing combination will yield an MMC diameter of 18.8 and a LMC diameter of 21.2. The flattened surface is 5.6 maximum with a positional limit of 2.9 diameter resulting in a minimum wall of 17.95.

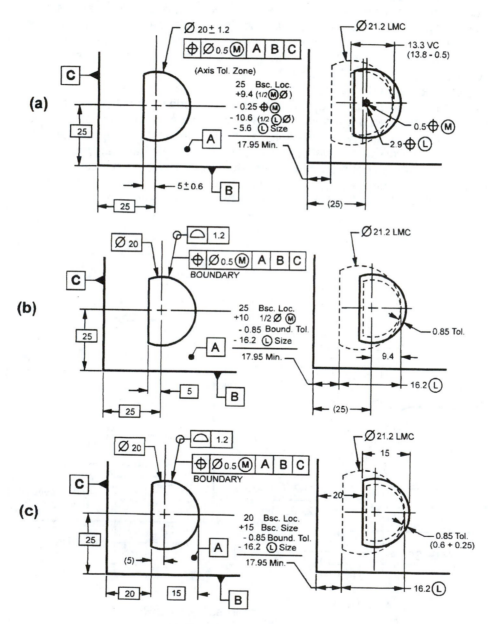

Figure 5-14 Boundary concept and other dimensioning.

Figure 5-14b illustrates a basic 20 diameter with a profile tolerance control for size and form, plus a positional tolerance control for location of the feature *boundary*. As shown, these combined controls allow the same MMC and LMC limits and minimum wall as did Figure 5-14a. Figure 5-14c illustrates a basic 20 diameter and profile tolerance control, but the position boundary limit is established by the feature surface. With this dimensioning and tolerancing combination, the MMC and LMC limits are the same as Figure 5-14a and b. The same resulting wall thickness of 17.95 is possible.

PROFILE CONTROLS SUMMARY

Profile controls:
May be applied with or without datums.
May be used in conjunction with size tolerances.
May be used to control irregular shapes.
May be applied as Composite controls.
May be used to control size, location, and form of coaxial shapes.
May be used to control coplanar relationships of two or more surfaces.
May be applied in combination with position controls, invoking the "Boundary" concept.
Is applied to features RFS, but datums may be RFS or MMC/LMC.

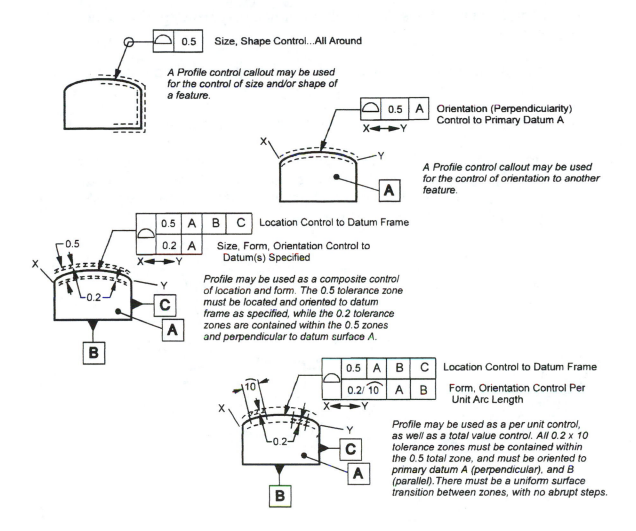

Figure 5-15 Profile controls applications.

EXERCISE 5-1 Profile

True or False, or fill in.

1. Profile is applied RFS unless stated otherwise. T   F

2. Profile tolerancing is always datum related. T   F

3. Profile may be used to control the _____ of flat surfaces.

4. Profile may be used to control _____ and _____ on cylindrical surfaces.

5. Profile may be used in combination with plus-minus or other geometric controls. T   F

6. In certain applications, the profile tolerance may extend beyond the size limits. T   F

7. Datums may be used to orient the profile control tolerance zone to other surfaces. T   F

8. The profile control tolerance zone is understood to be unilateral. T   F

9. In the figure below, indicate the *surface* between points X and Y to be relative to primary datum A and secondary datum B within 0.2. <u>Draw the tolerance zone</u>. Also indicate the same surface to have a profile tolerance of 0.1 between X and Y.

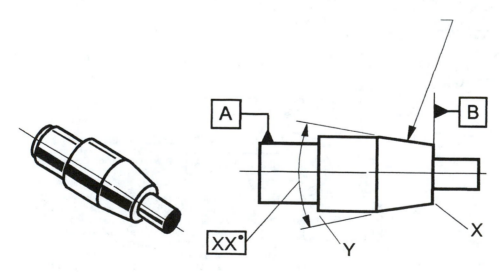

EXERCISE 5-2 Profile - Surface Definition

Complete the drawing control callouts 1 and 2 to satisfy or complete the requirements indicated.

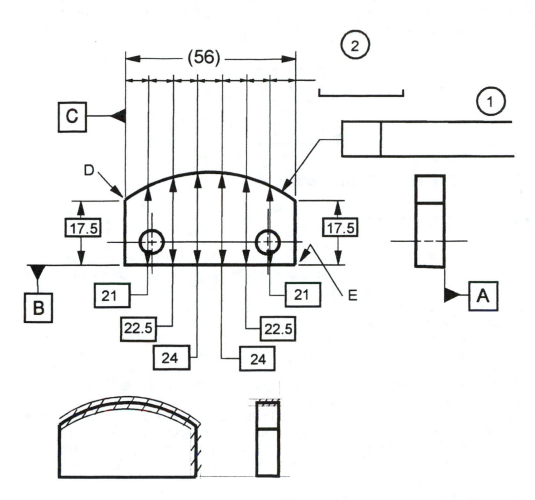

The surface between points D and E must lie between two profile boundaries 0.25 apart, perpendicular to datum plane A, equally disposed about the defined profile, and located with respect to datum planes B and C.

6   RUNOUT

Runout is a composite tolerance control, capturing all errors of size, form, orientation, and location, used to specify the functional relationship of one or more features of a part to a datum feature axis. See Figure 6-1. A runout tolerance indicates the permissible error of the controlled feature surface when rotated about the datum axis. The specified tolerance reflects the maximum *full indicator movement* (FIM) when the part feature is rotated 360 degrees. Runout is applied RFS only to both features and datums. It can be applied to a single surface element or to the entire surface. See Figures 6-2 and 6-3.

Figure 6-1 Runout controls.

CIRCULAR AND TOTAL RUNOUT

Circular runout is commonly understood, while total runout is generally described as containing enough readings to give reasonable assurance the surface has met the requirements specified. It is also a common understanding that *total* surface verification is not possible, with measurement tolerances, clearances, blends, and other factors involved.

With total runout, the indicator device must have the ability to traverse the surface, perpendicular or parallel to the primary datum specified. At times, the indicator is required to follow the contour of a curved surface to satisfy the requirement specified. This means that the device must travel normal to the surface measured, a difficult task at best. For these reasons, many feel that total runout (as well as other *total* controls) should be avoided and the use of multiple controls, in combination with circular runout, be employed instead. Nonetheless, the Y14.5 standard as well as the ISO standard offer total runout as a specification control that may be used on engineering drawings, and the ability to measure is naturally a consideration. The precise technique used to evaluate a total runout requirement is a quality engineering decision, normally based on past performance data, tool capability, risk, and other established quality standards.

Circular runout is a single element reading as illustrated by Figure 6-2a. The primary datum reference is normally a feature axis. All measurements are taken normal to the surface measured. Any circular reading may be used, as determined by the quality engineer, if not specified on the drawing. Figure 6-2b shows the potential effects of reversing the datum precedence (inadvertently or intentionally). Measurement results could be different on identical parts.

Figure 6-3a illustrates the total runout requirement and the necessity of moving the indicator to evaluate the surface. Measurements are taken normal to the surface. Figure 6-3b further illustrates the need to evaluate datum features to the common axis of a shaft. Opposite ends of the shaft have been designated as datums A and B, but additionally, the datums themselves must also be evaluated to properly review all elements of the shaft design. It should be obvious that datum error, if not controlled and measured, could completely nullify the readings of other features relative to the uncontrolled or erroneous framework.

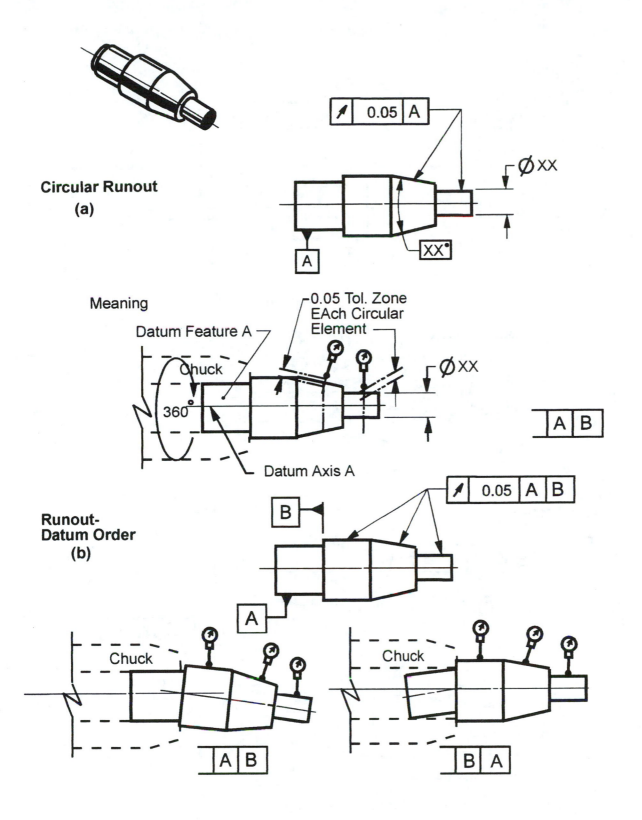

Circular Runout
(a)

↗ | 0.05 | A

⌀ XX

A

XX°

Meaning

0.05 Tol. Zone
EAch Circular
Element

Datum Feature A

Chuck

360°

⌀ XX

Datum Axis A

A | B

Runout-
Datum Order
(b)

B

↗ | 0.05 | A | B

A

Chuck

Chuck

A | B

B | A

**Figure 6-2 Circular runout and
datum order.**

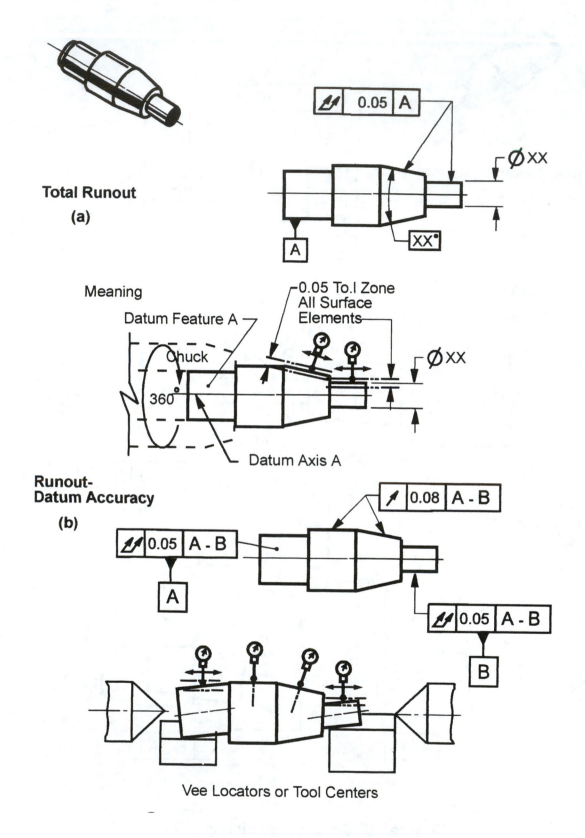

Total Runout

(a)

0.05 | A

Ø xx

XX°

A

Meaning

0.05 To.l Zone
All Surface
Elements

Datum Feature A

Chuck

360°

Ø xx

Datum Axis A

Runout-
Datum Accuracy

(b)

0.08 | A - B

0.05 | A - B

A

0.05 | A - B

B

Vee Locators or Tool Centers

Figure 6-3 Circular runout and
datum accuracy.

144

Figures 6-4 and 6-5 illustrate possible setups for measurement, along with the datum terms previously used. In Figure 6-4, notice that the shaft ends (tool centers) were used as primary datums A-B. In this example, the secondary datum is automatically established but not identified, because no lateral movement (float) can occur. The part is locked in the measurement fixture; therefore, all readings are direct or actual values.

Figure 6-5 shows the datums as A-B primary, with surface C as the secondary datum. This setup will allow lateral movement, which could influence the design integrity as well as measurement repeatability and part acceptance.

Looking further at Figure 6-6, we see the effects of using a shoulder as a datum, or stop. If the secondary datum feature C is not perfectly perpendicular to the primary datum features axis, lateral float can occur and can be read in the FIM of surface A, as well as other vertical surfaces. This condition is exaggerated in Figure 6-6b and c to illustrate the impact of the location C stop on FIM readings. If the stop can be placed at the axis as shown in Figure 6-6b and c, the readings may be taken as actual, because no datum float can occur or otherwise impact the readings. If, however, the stop is located away from the axis, at the edge of a shoulder, the readings may potentially double due to datum float. The "good" part features may be ruled as *out of conformance. When this condition exists, the readings should be halved.* Datum error must be considered when evaluating other features relative to conformance. Secondary datum error can not be avoided and therefore must be minimized or allowed for otherwise.

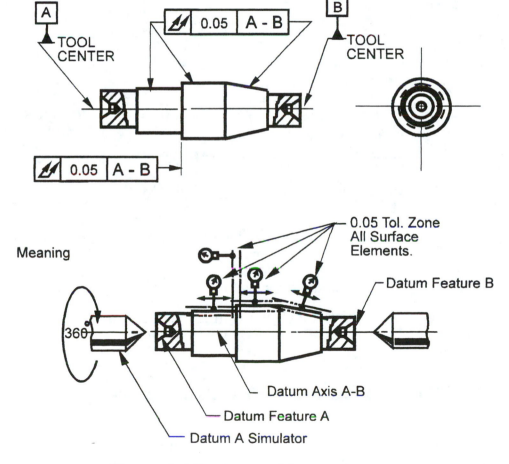

Figure 6-4 Datum machine centers.

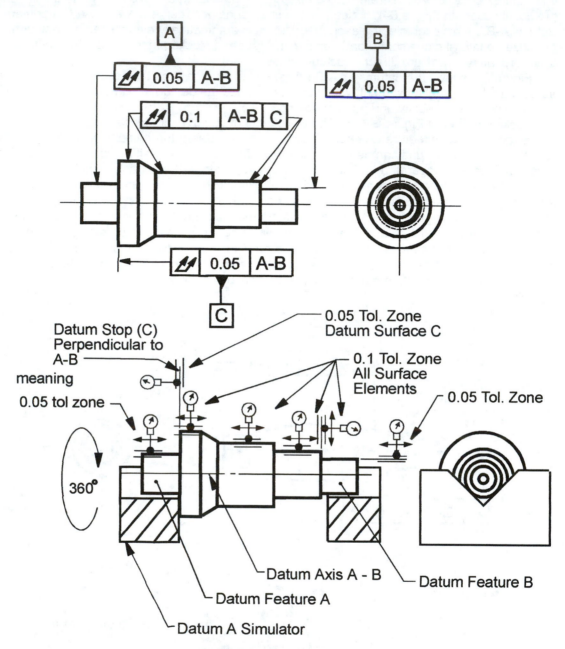

Figure 6-5   Datums: two functional diameters.

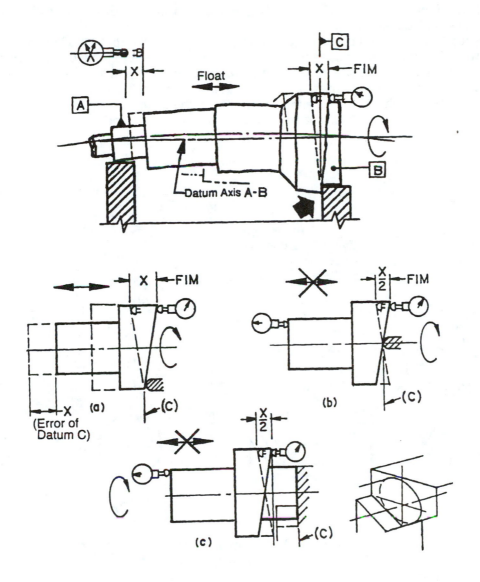

Figure 6-6 Secondary datum error.

147

VEE LOCATORS AND DATUM ERROR

From the previous discussion, it is obvious that when evaluating runout, one must use care to first find and resolve any error that may exist in the datum feature. Datum features can have size, form, and location error; therefore, if vee locators are used in the measurement setup, it is possible to have a perfectly concentric feature, read the ovality produced by the datum, and reject the part for incorrect reasons. Further, the location of the indicators can give varied results and in addition, the included angle of the vee locators may impact the measured results. These effects are shown in Figures 6-7 and 6-8.

We may read very little vertical error but considerable horizontal error with 90 degree vee blocks, but with 120 degree vee blocks, the vertical and horizontal errors tend to even out. These illustrations point out the importance of knowing and understanding the setup methods used, indicator locations, vee locator included angle, and datum size/shape when evaluating measurement results relative to runout and vee locators for the datum simulators. Further note that the vee locator angle may mask measured error on shafts with an odd number of lobes.

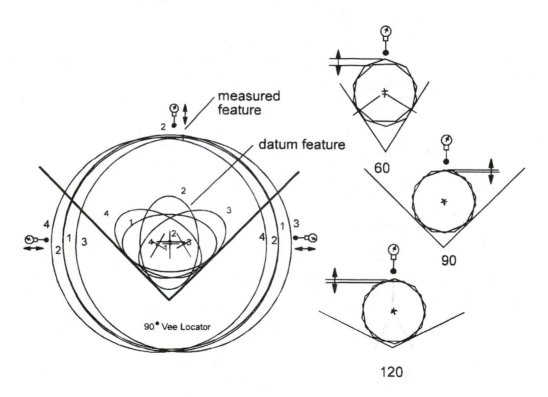

Figure 6-7 Datum ovality and measured error with 90 degree vee locators.

PRIMARY DATUM SURFACE

Occasionally, designs may require features be controlled to a primary surface datum using the diameter as a secondary datum. Figure 6-9 illustrates a collar diameter that is controlled to datum framework C, D. The primary design influence is the mounting surface C, with the bearing diameter D perpendicular to C. Carefully consider the effects shown, because measured error could be dramatically affected if the datums were reversed. The *secondary/tertiary datum rule* applies here, and any error will be reflected in the readings for the secondary datum feature.

148

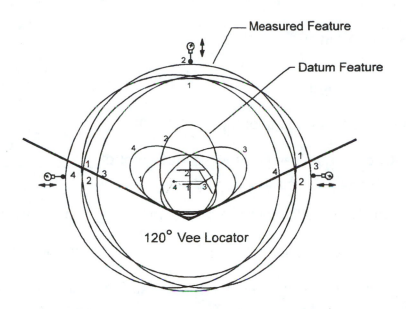

Figure 6-8 Datum ovality and measured error with 120 degree vee locators.

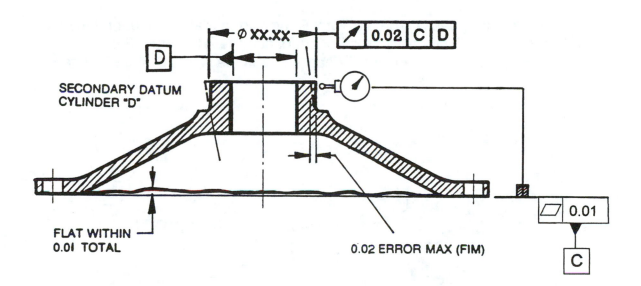

Figure 6-9   Datum mounting surface vs diameter.

RUNOUT SUMMARY

Runout controls apply to features RFS relative to datums RFS.
Runout controls are composite surface controls and contain all errors of size, form, orientation, and location.
Runout controls for a feature are considered relative to a datum axis.
Total runout controls apply to entire feature surfaces.

EXERCISE 6 Runout

True or False, or fill in.

1. Runout is understood to apply at MMC. T   F

2. Both Circular and Total Runout callouts control _____elements and include errors of _____and_____.

3. Additionally, Total Runout can control the _____on a cylindrical surface.

4. Runout may be applied with or without datum references. T   F

5. Runout tolerances are read at a feature surface, relative to a datum axis.
    T   F

6. The control of a convex or concave *surface* to a datum axis can be accomplished with circular runout. T   F

7. In the figure below, indicate Total Runout for surface X of 0.2 relative to datum A. Indicate a Circular Runout of surface Z to datum B of 0.1. Indicate a Total Runout of surface Y to datums A and B of 0.05. <u>Draw the tolerance zones</u>.

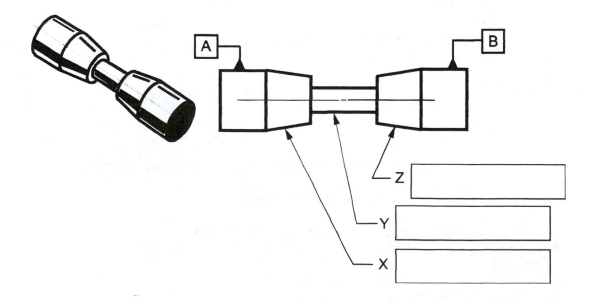

7   LOCATION CONTROLS

The three *location controls* are *position, concentricity* and *symmetry*. See Figure 7-1. These controls generally deal with features that have size and an axis or centerplane. Position controls may be applied MMC, LMC, or RFS, whereas concentricity and symmetry controls are applied only RFS. RFS is always implied, unless stated otherwise. We will start with *position*, the most commonly used location control and then cover *concentricity* and *symmetry*.

Position   **Concentricity**   **Symmetry**

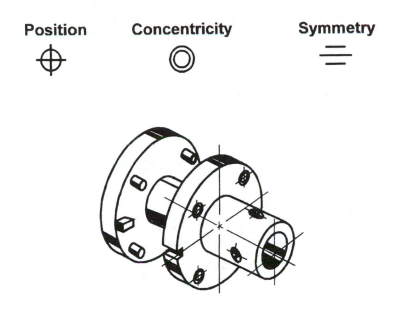

Figure 7-1 Location tolerance controls

POSITION

True position is a term used to describe the perfect (or exact) location of a feature point, axis, or plane (normally the center) of a feature in relation to a datum reference, or other feature. See Figure 7-2. The position tolerance is the total permissible variation in the location of a feature from its *true position*. For cylindrical features such as holes or bosses, the positional tolerance zone is a cylinder through the depth/length of the feature within which the axis of the feature must lie. For other features, such as slots or tabs, the Position tolerance zone is the total width between two parallel planes through the depth/length of the feature within which the centerplane of the feature must lie. As discussed in Chapter 5, it is also possible to combine *position* and *profile* tolerancing to relate a location tolerance to a feature *boundary*, if so specified.

A feature hole is rarely produced at a perfect or exact location. To repeat an exact hole location is even more difficult. Figure 7-2a illustrates possible locations of a single hole produced repeatedly in a loose fixture. As most parts contain patterns of holes, each hole will have a location tolerance, as shown in Figure 7-2b. Each intersection represents a true center, and each square represents a +/- .005 tolerance for each hole.

In addition to the individual hole tolerance, the pattern of four holes must be associated with a location control relative to the part outline, edges, or other functional features such as an axis or centerplane. Figure 7-3 illustrates the possible combined effects of +/-.005 diameter hole tolerance, along with +/-.015 pattern location tolerance from the part edges.

Figure 7-3 further illustrates one interpretation of hole-to-hole tolerance and pattern tolerance using the coordinate dimensioning system.

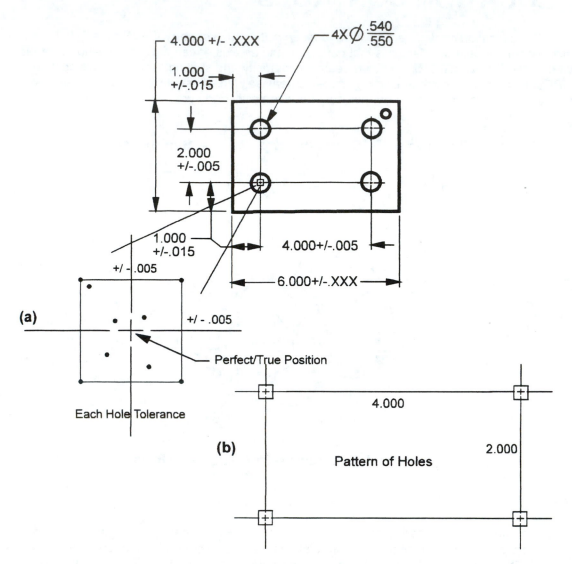

Figure 7-2 **Perfect true position**.

Position Tolerance Theory

The *position tolerance* theory is not complex, as illustrated by Figure 7-4. By converting a common tolerance, such as +/ - .005, that may appear on a coordinately toleranced drawing to positional tolerances, we see that we gain 57% tolerance area for manufacturing, with no sacrifice in the quality level. The shaded area was not available to manufacturing with the plus-minus system, however, so features produced at the corners of the +/ - tolerance zones are actually .007 from the theoretical perfect location. This converts to a tolerance zone diameter of .014. Further, with *positional tolerancing*, it is possible to have an MMC bonus tolerance, which is also unavailable with the plus-minus system.

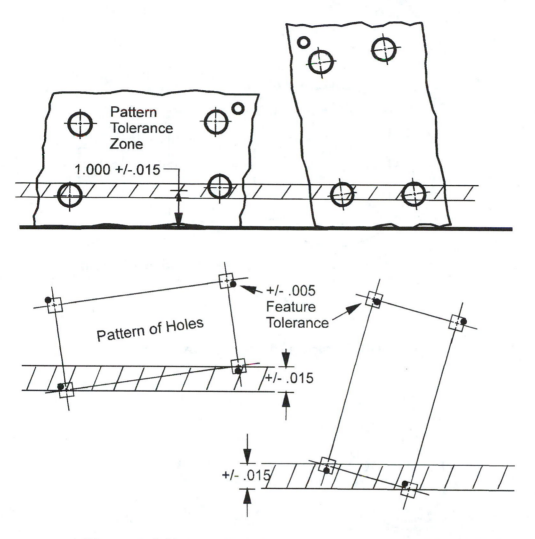

Figure 7-3 Hole pattern tolerance zones (see Figure 7-2).

The bonus tolerance exists when the feature hole(s) is larger (or shaft smaller) than its MMC size. For each unit of increased size of the feature, the same amount can be added to the position tolerance. As noted in the general tolerancing section, three elements affect mating part fits: hole size, fastener size and total tolerance. See Figure 7-5. Mathematically, these elements work as a unit:

$$T = H - F \qquad H = F + T \qquad F = H - T$$

where T = tolerance, H = hole size and F = fastener size

If one element changes, the other elements are affected. Further, the perpendicularity of the feature holes relative to the primary datum must be accounted for in calculating tolerances. This is accomplished by the tolerance zone rule, "Tolerance zones exist for full feature length, width, or depth."

153

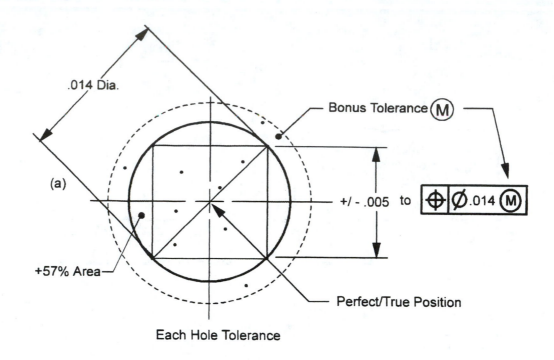

Figure 7-4 Position tolerance theory.

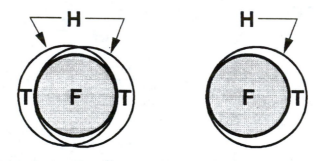

Figure 7-5 Tolerance elements.

Position Tolerance and Gaging Formulas

Floating fastener condition   T = H - F   Clearance holes, all parts.

Fixed fastener condition $T = \dfrac{H - F}{2}$ Threaded or close fit holes, all parts.

Gage element mize for male feature   G = feature + T

Gage element size for female feature   G = feature - T

154

Figure 7-6a illustrates a simple four-hole pattern located by standard rectangular coordinate dimensioning. Using this figure as an example, we can convert the dimensioning to GDT positional dimensioning and tolerancing. We are not redesigning the part, only converting the dimensioning and tolerancing. In Figure 7-6b, the hole size been converted using the formula of Figure 7-4, and the location tolerance of +/- .005 has been converted to a diameter zone of .014. The fastener is a .500 diameter bolt MMC, and if we add the bolt size to the .014 tolerance, we have the clearance hole size of .514 dia. MMC. The feature control frame reads the true position of the four feature holes as .014 diameter perpendicular to primary datum A when the holes are at MMC size. Note that this is a *floating fastener* design in that all parts have clearance holes.

The next step in conversion concerns the hole pattern location relative to the datum framework. The pattern is located from the part edges, left edges, and lower edges as shown by the two 1.00 +/- .015 dimensions. These edges are the *secondary* and *tertiary datums* used to locate the pattern. Converting the +/- .015 tolerance, we have a diameter tolerance zone of .042 that is relative to the datum frame A, B, C. This control frame is placed above the feature control frame, thus giving us a *composite positional tolerance* control. The position symbols are entered once and are applied to both upper and lower entries. Figure 7-7a illustrates this step.

The last step in the process is the consideration of datum target points or the possible use of zero positional tolerance for the feature relationship. Datum targets would define the framework more precisely, as discussed. To use zero position tolerance, we must use MMC and make the clearance hole size the same as the fastener size (.500).

The maximum clearance hole size might be a standard oversize, such as .562 diameter. This would allow manufacturing the flexibility of choosing a tolerance by virtue of drill or punch size selection. An example may be choice of a 17/32 drill (.531), which would allow a .031 tolerance zone. With the above conversion completed, all available tolerance has been used, and the datum frame precisely defined, both requirements much asked for by manufacturing and suppliers. A comparison of both tolerance zones is shown in Figure 7-8.

MAXIMUM HOLE SIZE

The maximum hole size (.562) was pulled out of the air as a reasonable standard oversize. If another approach were taken, it may be possible to come up with a better engineered, defendable number. Consider screw thread designs. For thread class 2 standard applications, we can expect threads to be produced at about 55 to 65% effective thread form, with 70% effective threads considered near perfect for ferrous metals. Generally screw heads are considered to have a load-carrying capacity of 2X thread strength. If data from the *Machinery Handbook* industry standards are applied, we find the bearing area under the head of a 10 mm capscrew to be 15.6 mm at MMC. Subtracting the screw thread size of 10 mm gives a 5.6 mm total bearing area, half of which is 2.8 mm. The use of 70% of this number is a reasonable safety factor (taken from thread form applications) to arrive at 2 mm oversize for a maximum hole size of 12 mm for a 10 mm screw. Thus a formula 1.2 x fastener diameter MMC develops. This formula may be adequate in heavy industry and with ferrous metals but would need modification if used with softer metals, plastics, or other less dense materials. Still, the principle is that the formula could not be less than 1.0 x fastener size, and whatever ratio is used, we should avoid arbitrary selection of tolerances based on undefendable past practices or tradition. Tolerances should be justified by some engineering and mathematical logic. See Figure 7-9.

With this information, the holes in Figure 7-7b could have a maximum size of .600, provided that the hole size, with appropriate bonus tolerance (resultant condition), creates no design problems due to inadequate wall thickness at the part edges. Always consider the effects of LMC before finalizing design tolerances and callout controls. Holes may be produced with size, location and/or perpendicularity errors (or combinations). Figure 7-10 illustrates these errors applied to three holes, which may be repeated in any number of holes.

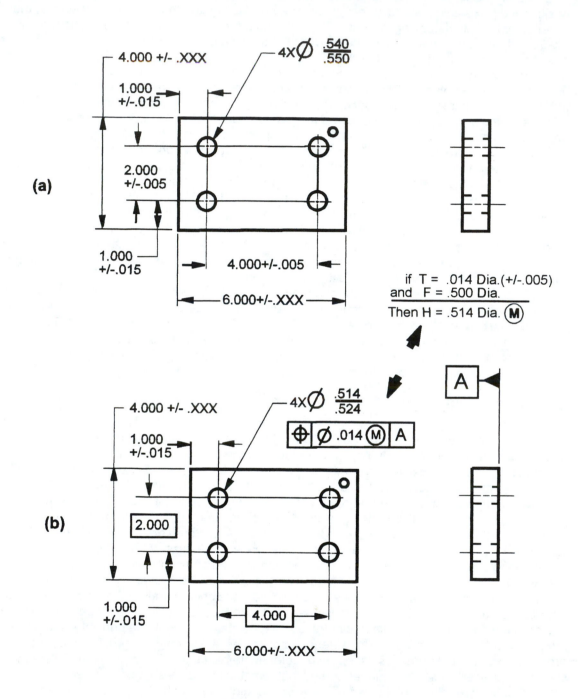

(a)

4xØ .540/.550

4.000 +/- .XXX

1.000 +/-.015

2.000 +/-.005

1.000 +/-.015

4.000+/-.005

6.000+/-.XXX

if T = .014 Dia.(+/-.005)
and F = .500 Dia.
Then H = .514 Dia. (M)

(b)

4xØ .514/.524

⊕ | Ø .014 (M) | A

4.000 +/- .XXX

1.000 +/-.015

2.000

1.000 +/-.015

4.000

6.000+/-.XXX

A

Figure 7-6 Rectangular pattern, coordinate tolerance conversion.

156

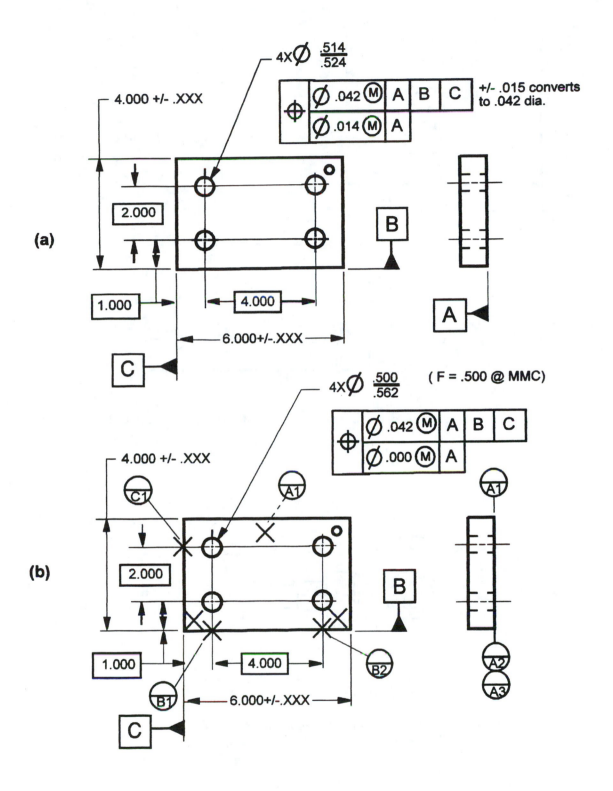

(a)

4X⌀ $\frac{.514}{.524}$

⊕	⌀ .042 Ⓜ	A	B	C
	⌀ .014 Ⓜ	A		

+/- .015 converts to .042 dia.

4.000 +/- .XXX

2.000

1.000

4.000

6.000 +/-.XXX

B

A

C

(b)

4X⌀ $\frac{.500}{.562}$ (F = .500 @ MMC)

⊕	⌀ .042 Ⓜ	A	B	C
	⌀ .000 Ⓜ	A		

4.000 +/- .XXX

C1

A1

2.000

1.000

4.000

B2

B1

6.000+/-.XXX

B

A1

A2

A3

C

Figure 7-7 Rectangular pattern, coordinate tolerance conversion.

157

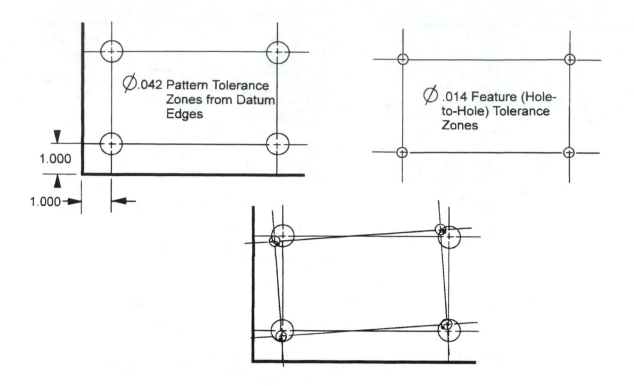

Figure 7-8   Composite position tolerance zones.

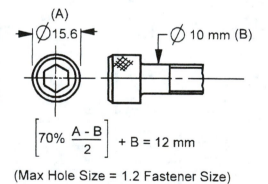

$$\left[70\% \; \frac{A - B}{2}\right] + B = 12 \text{ mm}$$

(Max Hole Size = 1.2 Fastener Size)

Note: This formula is typical for general applications
of UN class 2 threads in ferrous metals only.

Figure 7-9   Maximum hole size guideline.

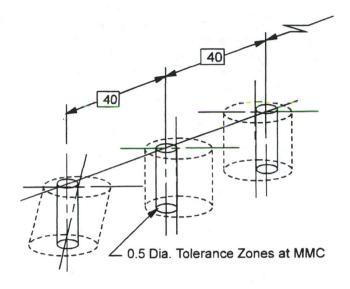

0.5 Dia. Tolerance Zones at MMC

Figure 7-10 Hole location, perpendicularity error.

TOOLING AND FIXTURING

By using Figure 7-7 along with the background established on datum target points, we can apply dimensioning and tolerancing principles to ensure design integrity is maintained throughout the manufacturing process. The designer should consider the impact of tolerances on the processing of product, including tooling/fixturing and inspection. This consideration is necessary to ensure interchangability, measurement repeatability, and compatibility of service replacement parts. Proper targeting and fixturing, along with correct tooling/gaging tolerances, will aid this process.

Figure 7-7b shows a datum framework with associated target points/lines. These targets may be used by manufacturing and inspection to achieve product integrity. The processing fixture for Figure 7-7 could look like that shown in Figure 7-11.

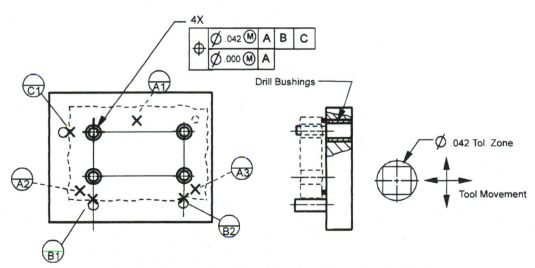

Figure 7-11   Tooling fixture for Figure 7-7.

Most tools used to generate holes in parts of this nature move in X and Y directions, thus creating a square tolerance zone, rather than a circular tolerance zone defined by the positional control callout on the drawing. The drawing control callout tolerance for the hole pattern is a diameter of .042 relative to datums A, B, and C. The tool engineer needs to convert the drawing tolerance to the appropriate manufacturing tolerance, as shown in Figure 7-12. This conversion is made on an RFS basis.

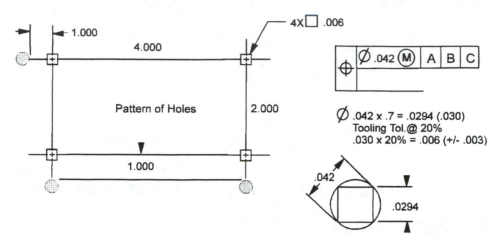

$\oslash$ .042 x .7 = .0294 (.030)
Tooling Tol.@ 20%
.030 x 20% = .006 (+/- .003)

Figure 7-12 Fixture tolerance calculations.

Going further, the lower control callout must be contained within the upper control limits. The lower control is .000 tolerance at MMC, depending on hole size. If we generate a hole size of .530 diameter (.030 oversize) and use the same process as above, the lower callout generates the following tolerances for the process.

The positon tolerance of .000 at MMC to datum A is:

.530 - .500 MMC = .030
.030 dia. x .7   = .021 round to .020
tooling tol.   = .020 x 20% = .004 (+/-.002)

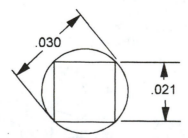

We can conclude that the hole-to-hole tolerance of the lower control callout requires tool accuracy of +/- .002 between drill spindles; this tolerance, however, is not related to any datum reference or fixture targets. The tolerance zones are perpendicular to surface A.

The gage designer and/or inspector may go through a similar process, except that the inspection tolerance allowed is 5% of the feature size tolerance (ANSI B4.4). In this example, two gages are required due to two datum frame references; one relative to the datum targeting framework, the other relative to surface A only and the hole-to hole feature relationships.

The following process will reject parts with line-to-line fit conditions when all variables are at the worst case.

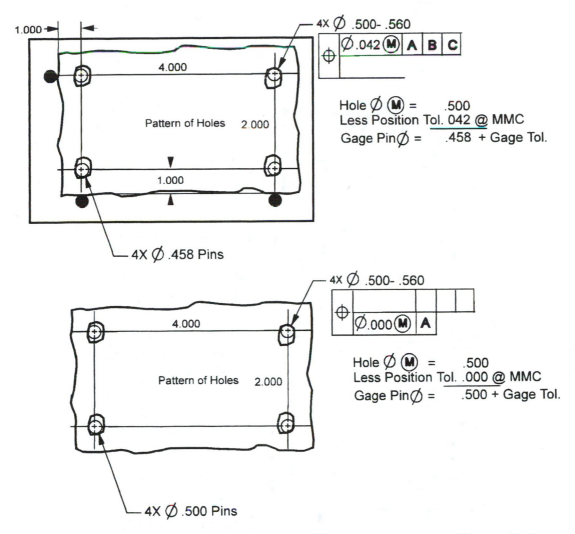

Hole $\emptyset$ (M) = .500
Less Position Tol. 042 @ MMC
Gage Pin$\emptyset$ = .458 + Gage Tol.

Hole $\emptyset$ (M) = .500
Less Position Tol. .000 @ MMC
Gage Pin$\emptyset$ = .500 + Gage Tol.

ANSI B4.4 (being replaced by ASME B89.3.6) allows the gage design a total of 10% of the feature size tolerance for measurement tolerances and allowances. As mentioned earlier, this tolerance is divided into 5% for the gage design and 5% for wear allowance. Further, the allowance is normally divided equally between form and location of the gage features. Refer to Figure 1-7 and the fundamental gaging principles of page 17. As gages are not always possible or practical for parts evaluation, *coordinate measuring machines* and other computer-assisted devices, as well as open setup surface plate inspection techniques, are often used. The measurement allowance is also available for these alternative methods.

This exercise illustrates the need for all areas dealing with engineering drawings in the product development though manufacturing process to be working from the same set of ground rules. If this or similar processes outlined in QS9001 along with existing National Standards are used, and if tolerances and processes defined in ASME Y14.5M, ANSI B4.4, B4.1, and B4.2 are followed, we will have gone a long way to ensure design integrity, product interchangability, and repeatability of measurement results. Service parts compliance will naturally follow.

Figures 7-13 and 7-14 illustrate simple gages for clearance holes. Using the process just completed, review the gages and gage pin sizes in these figures for compliance.

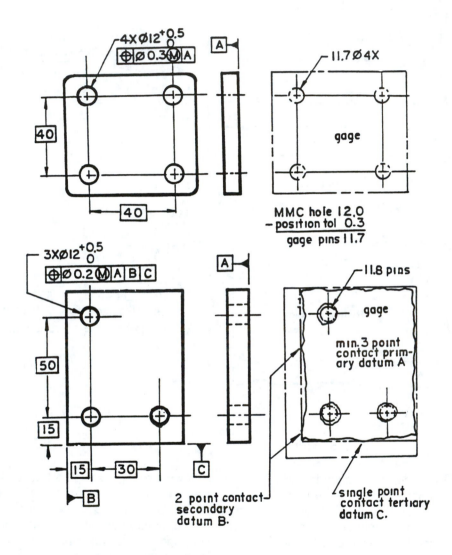

Figure 7-13 Simple functional receiver gages.

162

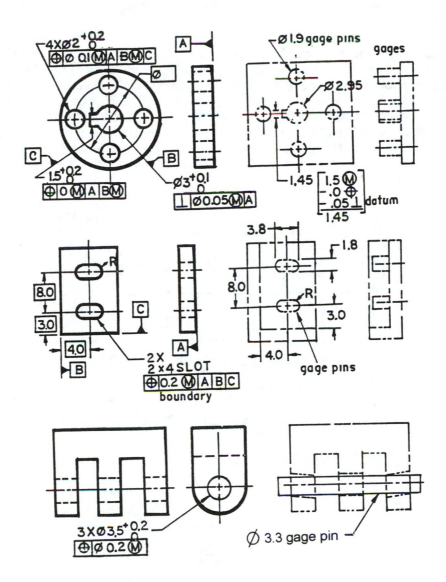

Figure 7-14 Simple functional receiver gages.

163

Applications of Position Controls

Let us explore a few applications where position tolerancing controls might be used. The illustrations shown are not the only possibilities for a particular type control, but may serve as options to consider.

Feature Symmetry Figure 7-15 shows two methods for locating a pattern of holes symmetrically in a part. Notice the location of the datum symbols B and C in Figure 7-15a. By placing the symbol as an extension of the 4.000 and 6.000 dimensions, we have specified the centerplanes of these features as the designated datums, RFS. The gaging or measurement system must be variable to allow for datum feature size tolerance. If MMC had been applied to the datums B and C, a fixed-type receiver gage could have been designed to accept or reject the part. Figure 7-15b illustrates the use of a minimum dimension (possible snap gage) to locate the pattern of holes from edges chosen, thus ensuring symmetry of the hole pattern in the part. Other techniques are possible.

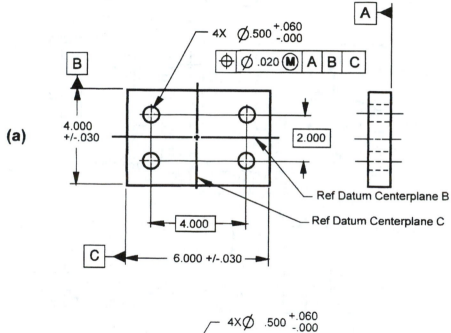

(a)

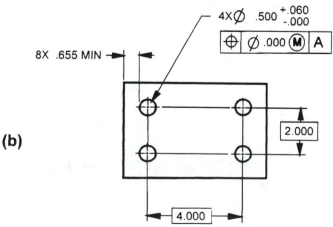

(b)

Figure 7-15 Hole patterns symmetrically located.

Multiple Patterns of Holes Figure 7-16 shows two sets of holes with the same datum references, in the same order, and with the same material conditions invoked. These control callouts imply that the two sets of holes may be considered as a single pattern for design and measurement. If a single implied set of two separate patterns were not the design intent, the words "SEPARATE REQUIREMENT" ("SEPT REQT") are added beneath the feature control frame, as shown in Figure 7-17.

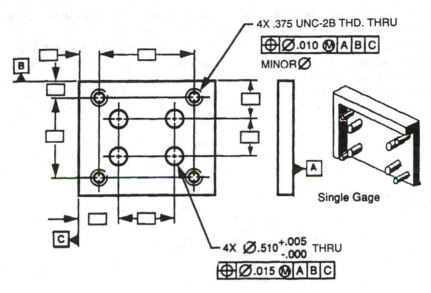

Figure 7-16 Multiple patterns, standard requirement.

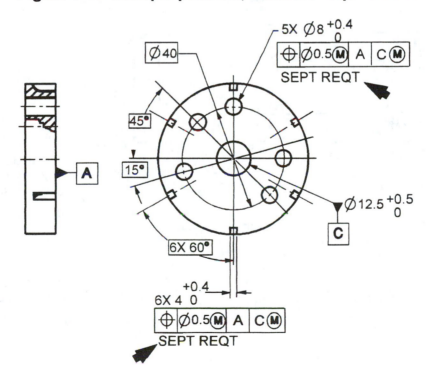

Figure 7-17   Multiple feature patterns, separate requirements

165

Multiple Datum Reference Frameworks At times, there may be a requirement to locate features from more than one datum framework. The threaded holes in Figure 7-16 are to located from the framework A, B, C. In addition, let's assume that we have a requirement to control the holes from other features as well. Figure 7-18a illustrates a composite position control with the threaded hole pattern controlled from both the part edges, as well as the pair of dowel locators. One requirement is from framework A, B, C, whereas the additional control is from the A, D framework. With these requirements, we have created additional restrictions, because the feature holes must reside simultaneously in both tolerance zones. To help sort out the tolerancing and gaging implications of Figure 7-18a, look at explanatory Figure 7-18b and c. Figure 7-18b shows allowable tolerance zones and a possible functional gage design for the two datum holes D and the four threaded feature holes, both relative to datum frame A, B, C. Because both sets of holes use the same framework as datum reference (A, B, C), these requirements could be reflected in a common gage, as shown. Figure 7-18c illustrates that the two datum D holes could be separately verified for compliance with the control .002 dia. MMC relative to datum surface A, with the four threaded holes relative to surface A and secondary datum holes D, RFS. Because the datum holes are expressed RFS, the gage pins need to be variable and adjustable. There is no relationship to the A, B, C framework in this requirement.

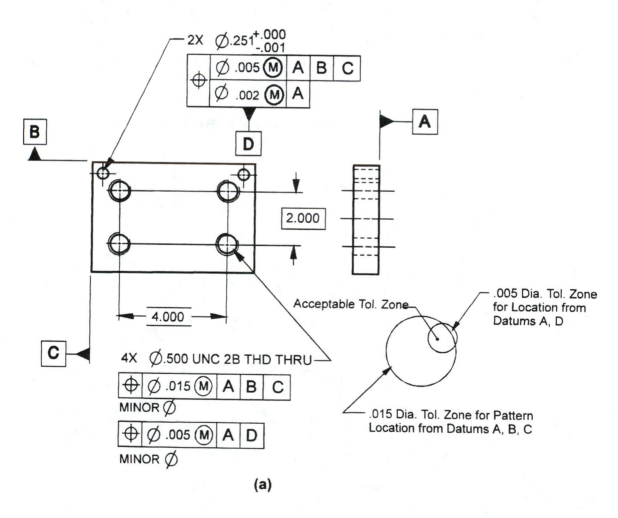

(a)

Figure 7-18 Multiple datum drameworks and controls.

166

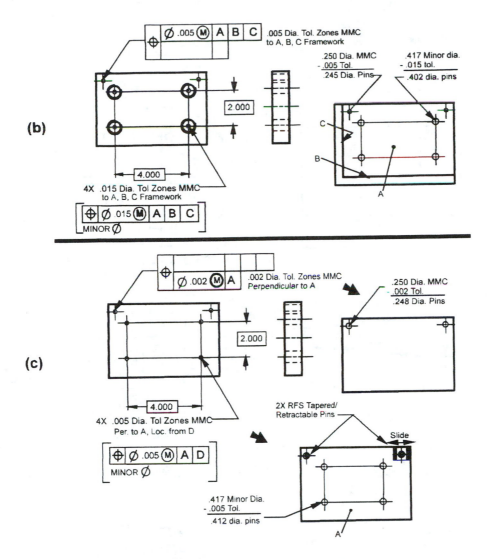

(b)

Ø .005 Ⓜ A B C   .005 Dia. Tol. Zones MMC to A, B, C Framework

.250 Dia. MMC
- .005 Tol.
.245 Dia. Pins

.417 Minor dia.
- .015 tol.
.402 dia. pins

2.000

4.000

4X .015 Dia. Tol Zones MMC
to A, B, C Framework

⊕ Ø .015 Ⓜ A B C
MINOR Ø

(c)

Ø .002 Ⓜ A   .002 Dia. Tol. Zones MMC
Perpendicular to A

.250 Dia. MMC
.002 Tol.
.248 Dia. Pins

2.000

4.000

4X .005 Dia. Tol Zones MMC
Per. to A, Loc. from D

⊕ Ø .005 Ⓜ A D
MINOR Ø

2X RFS Tapered/
Retractable Pins

Slide

.417 Minor Dia.
- .005 Tol.
.412 dia. pins

Figure 7-18 (continued)

Conical Tolerance Zones Very long shafts and spacers may make it difficult to hold a tolerance through the full depth or length of a feature, per the fundamental rule for tolerance zones. Drill run, material composition inconsistencies, or other problems may create a need for tolerance zones larger at a feature exit than at the origin or entry. The controls applied in Figure 7-19 are one solution to this condition. This control callout provides a 0.5 diameter tolerance zone at entry, datum surface B, but with a 1.0 dia. tolerance zone at the exit, surface C. Care should be taken to ensure that this technique does not result in insufficient wall thickness at surface C. Also, if applied MMC as shown, and if the feature were made to LMC size limits of 12.5 diameter, a bonus location tolerance to add to the potential thin wall problem will result. Apply this technique carefully!

Radial Hole Patterns Radial hole patterns are common in cylindrical parts. Often, the primary datum might be an axis, as shown in Figure 7-20. This figure shows a requirement at MMC that would allow a functional gage to be used. The gage would pass through datum diameter A and stop on datum surface B. Six gage pins would pass through the feature holes and into the gage. The depth of the holes in the gage would be equivalent to the wall thickness of the part.

167

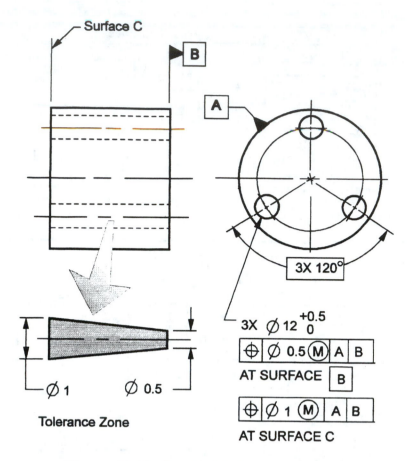

Figure 7-19 Conical tolerance zone.

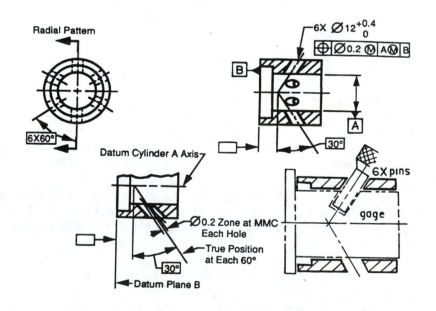

Figure 7-20 Radial hole patterns.

Multiple Operations It is common for more than one feature to share the same axis or centerplane, such as holes and counterbores or counterdrilled holes. If the features control callout appears at the end of a series of operations in a note format, the implication is the control specified applies to *all operations* preceding the callout. If this is not the design intent, separate control frames should be specified, as illustrated in Figure 7-21a.

Individual Holes as Datums Some designs require the use of shoulder screws or ring dowels to align components of an assembly. These designs may require each individual hole used as a datum for succeeding operations. When this requirement exists, the control callouts in Figure 7-21b should be considered. The illustration shows the hole tolerance zone is zero MMC, however other tolerancing methods may be used. With shoulder screws, we have a fixed fastener condition. The tolerancing formula is $T = \dfrac{H - F}{2}$.

Figure 7-21 Counterbores: (a) multiple operations;
(b) individual datum holes.

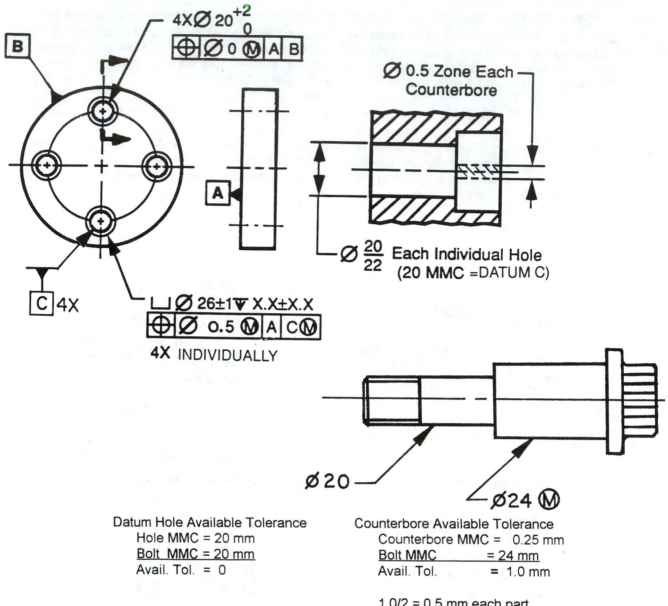

4XØ 20 $^{+2}_{0}$

⊕ | Ø 0 Ⓜ | A | B

Ø 0.5 Zone Each Counterbore

Ø $\frac{20}{22}$ Each Individual Hole
(20 MMC =DATUM C)

⌴ Ø 26±1▼ X.X±X.X

⊕ | Ø 0.5 Ⓜ | A | C Ⓜ

4X INDIVIDUALLY

C 4X

B

A

Ø 20

Ø24 Ⓜ

Datum Hole Available Tolerance
Hole MMC = 20 mm
Bolt MMC = 20 mm
Avail. Tol. = 0

Counterbore Available Tolerance
Counterbore MMC =   0.25 mm
Bolt MMC   = 24 mm
Avail. Tol.   = 1.0 mm

1.0/2 = 0.5 mm each part

(b)

Figure 7-21 (continued)

Other Than Cylindrical Tolerance Zones Figure 7-22a illustrates the use of position tolerance bidirectionally to develop a banana-shaped tolerance zone. When applying this type control in this manner, the angle dimensions and other related dimensions must be basic.

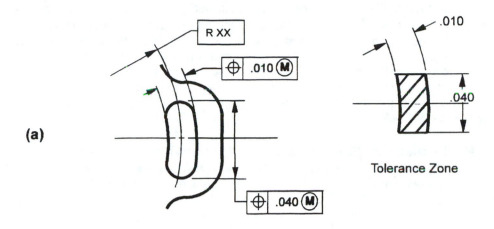

(a)

Tolerance Zone

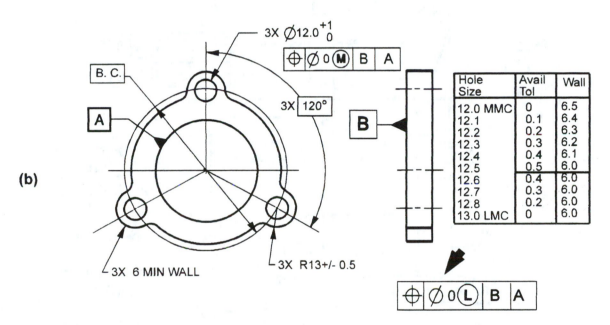

(b)

Hole Size	Avail Tol	Wall
12.0 MMC	0	6.5
12.1	0.1	6.4
12.2	0.2	6.3
12.3	0.3	6.2
12.4	0.4	6.1
12.5	0.5	6.0
12.6	0.4	6.0
12.7	0.3	6.0
12.8	0.2	6.0
13.0 LMC	0	6.0

**Figure 7-22 (a) Other tolerance zones;
(b) restricted tolerance zones.**

Restricted Zones Figure 7-22b illustrates one method of restricting a tolerance zone to avoid thin wall conditions at the three bosses. The principles of MMC bonus tolerance could be a problem with this example. In this figure, if the holes were produced to the maximum limits of size (LMC 13 mm),the position tolerance would be 1.0 mm. This combination of the largest hole size and bonus position tolerance could result in inadequate wall thickness, a resultant condition. One method of avoiding this condition is with the use of a note, as shown in the figure. The use of LMC in the control frame is also an option. As the holes in Figure 7-22b depart from their MMC size, at about 12.6 mm, they must return to the true location to maintain the 6.0 minimum wall.

Bidirectional Tolerance Zones The technique illustrated in Figure 7-23 creates a bidirectional, or rectangular, tolerance zone. In the feature control frames, note the absence of a diameter symbol. Because round gage pins in round holes do not generate rectangular tolerances, designing the tolerance into the gage plate with square shoulder pins is an option. The tolerance is added to the pin shoulder size, which then becomes the size of the rectangular hole in the gage plate. The rails of the gage simulate the secondary and tertiary datum edges B and C. When the part is correctly located on the gage, the gage pins must pass through the gage plate and through the .500 MMC holes.

Slots (Elongated Holes) Tolerance Zones Slotted holes may incorporate the *boundary principle,* which establishes a tolerance zone boundary when using positional tolerance controls. Positon controls are normally applied to feature axes or centerplanes. By using the boundary principle in Figure 7-24a, a functional gage design is possible. The tolerance zone exists outside the boundary created by the MMC shape of the slot. Figure 7-24b illustrates one additional method of dimensioning slots, and ASME Y14.5 provides other methods.

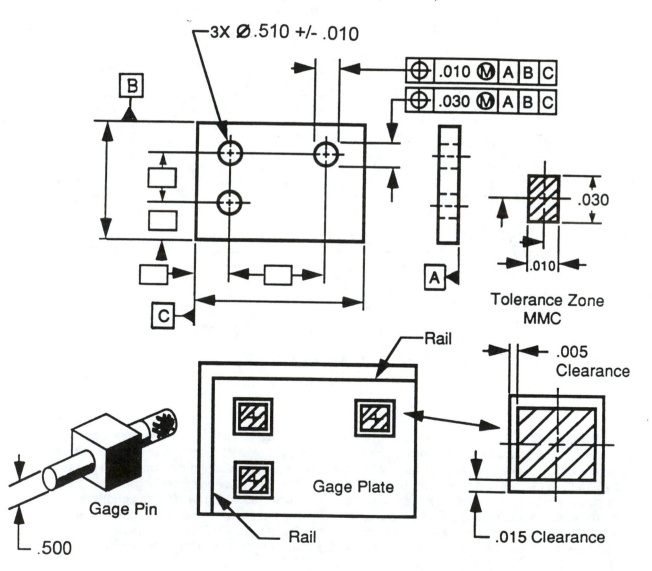

Figure 7-23 Bidirectional tolerance zones.

If the boundary principle is not invoked and the word "BOUNDARY" is not used, the tolerance zone exists at the feature axis or centerplane, at the intersection of basic dimensions, and would be rectangular, as illustrated by Figure 7-24b. Functional gaging would still be possible, with the tolerance zone designed into the gage plate, as shown previously in Figure 7-23.

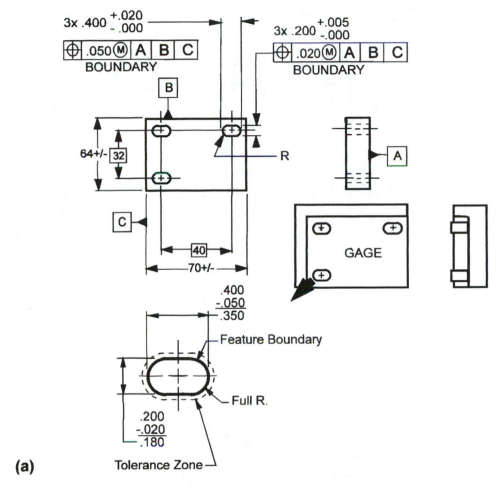

Figure 7-24   Boundary principle. (a) Elongated holes (slots).

Composite Positional Tolerance

The term *composite positional tolerance* has been used, but it has not been thoroughly explored. Let us recall Figure 7-7. Composite tolerancing provides a combined application of positional tolerancing for the location of feature patterns as well as the interrelationship of features within these patterns. The position symbol is entered once and is applicable to both upper and lower entries of the control frame. Each entry is complete and may be separately verifiable. Figure 7-25 shows a composite position control frame.

Datum references specified in the lower entry govern the feature-to-feature relationship and the orientation of the pattern of features to the specified datums. One or more datum specified in the upper control callout are repeated, as applicable, and used in the same order of precedence. Figure 7-26 illustrates a composite position control.

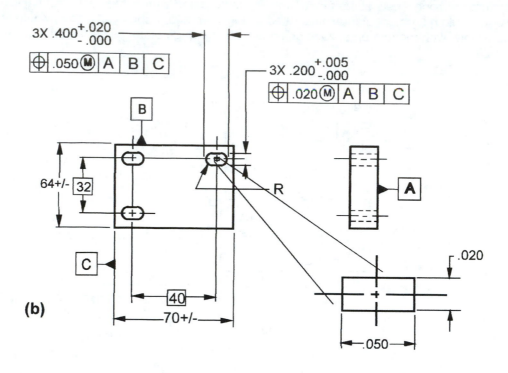

Tolerance Zone MMC

(b)

Figure 7-24 (b) rectangular tolerance zones.

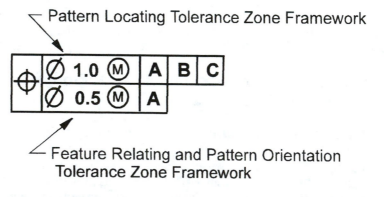

Pattern Locating Tolerance Zone Framework

Feature Relating and Pattern Orientation
Tolerance Zone Framework

Figure 7-25 Composite positional tolerancing.

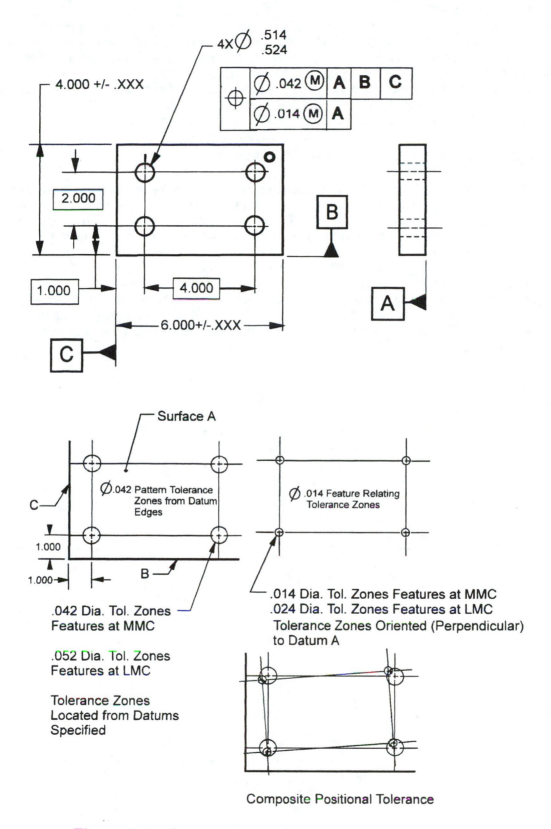

Figure 7-26 Composite position control example.

The upper portion of the control callout specifies the four-hole pattern to be located and oriented relative to the A, B, C framework. That is, it must be perpendicular to A primary and located from secondary B and tertiary C within a .042 diameter tolerance zone when the features are at MMC size. The lower control callout specifies that the individual holes are relative to one another and perpendicular to surface A within a .014 diameter tolerance zone; it also specifies when the features are at MMC size. The lower control location tolerance must fall within the upper control zones. Two gages would be required for this composite control.

Composite Position Control Applications

Figures 7-27 and 7-28 illustrate the use of composite position controls for alignment of groups of holes. As before, the upper control establishes limits for the pattern, as a group, relative to the datum framework. The lower control establishes the hole-to-hole relationship and the orientation to the datums specified as they are invoked. To refine the orientation of the feature relating tolerance zones of the lower control, datum references are repeated as necessary and applicable. With composite position tolerancing, it is important to remember that the *basic* dimensions that control the pattern of features to the datum framework apply to the upper control only; the lower callout determines orientation to specific datum references and hole-to-hole locations. Should it be required that both location and orientation be controlled to varied datum references, *multiple single segment* controls should be used as illustrated in Figure 7-39, later in this chapter.

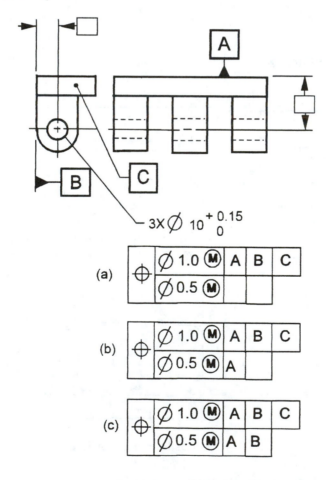

Figure 7-27 Composite position controls: aligned holes.

176

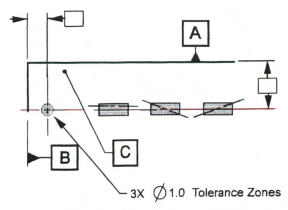

3X ⌀1.0 Tolerance Zones

For the Upper Control Callout
The 1.0 diameter tolerance zones must be aligned and
basically located from the datum framework A, B, C.

For the Lower Control Callouts

(a)

⊕	⌀1.0 Ⓜ	A	B	C
	⌀0.5 Ⓜ			

(b)

⊕	⌀1.0 Ⓜ	A	B	C
	⌀0.5 Ⓜ	A		

(c)

⊕	⌀1.0 Ⓜ	A	B	C
	⌀0.5 Ⓜ	A	B	

A

3 X 1.0
CyL. Zones

B

The 3 x 0.5 diameter zones
are aligned, must lie within the
1.0 diameter zones, but are not
controlled to any datums.

A

3 X 1.0
Cyl. Zones

Parallel

B

The 3 x 0.5 diameter zones
are aligned, must lie within
the 1.0 diameter zones, and
must be parallel to datum B.

A

3 X 1.0
Cyl. Zones

Parallel

B

Parallel

The 3 x 0.5 diameter zones
are aligned, must lie within the
1.0 diameter zones, and must
be parallel to datums A and B.

**Figure 7-28 Upper and lower controls of
Figure 7-27 explained.**

Note that the composite principle also applies to profile tolerancing as described in Chapter 5. Comparisons of both *composite position tolerancing* and *single segment position tolerancing* are illustrated later, which should help in determining which controls to use in the design.

The upper control of Figure 7-27 should be clear. The lower control callout of Figure 7-7a has no datums invoked. There exists full freedom of movement of the 0.5 tolerance zones, as a group, within the upper control tolerance zones. See Figure 7-28a. In Figure 7-27b, datum A has been invoked, removing one degree of freedom; therefore, the three 0.5 tolerance zones must be aligned and parallel to datum A only. See Figure 7-28b. In Figure 7-27c, datums A and B have been invoked, meaning that the lower control 0.5 tolerance zones must be parallel to both datums A and B. See Figure 7-28c.

Circular Patterns Figure 7-29 illustrates the use of composite positional tolerancing for a circular pattern of holes located from a primary datum surface G and a secondary datum diameter H, MMC. The principles applied are the same as for aligned holes. The upper control applies to the pattern of holes located from the specified datum framework, and the lower control callout applies to the hole-to-hole relationship and the orientation of the pattern to the datums specified. In this case, the Pattern location is from datums G and H, whereas the feature relationship and orientation is relative only to datum surface A.

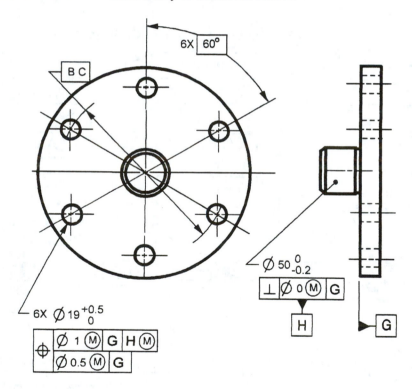

Figure 7-29   Composite position tolerance: circular pattern.

By using the previously noted gaging formulas, the gage pins for the upper control would be 18.0 mm (19.0 MMC - 1.0 tol.). The pins must be relative to datums G (perpendicular) and H (location). The part must fit over the gage with datum pin G at its virtual condition MMC size of 50 mm. The part must rest, of its own weight, flush with primary surface G. Feature gage pins may be fixed or loose, but must penetrate the full feature hole depth.

178

A second gage is required, or one made in combination with the first gage, with feature gage pins of 18.5 diameter (19.0 - 0.5), but without a centering datum hole H. Because datum H is not invoked in the lower control, it must not be part of the gage design, which might be accomplished with a removable insert. As can be seen from Figure 7-30, this second gage requires closer control of the hole-to-hole relationship and perpendicularity to primary datum surface G, but without the location control to secondary datum diameter H.

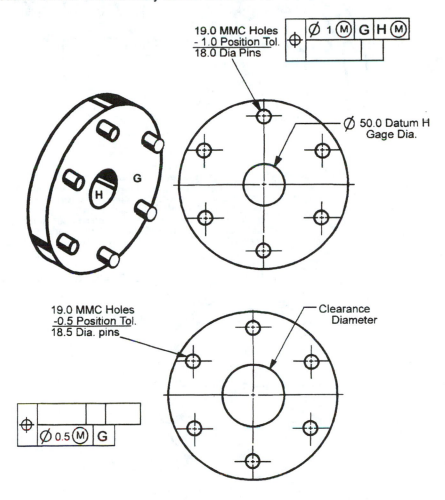

Figure 7-30   Possible gage(s) for Figure 7-29; part at MMC.

Figure 7-31a illustrates the same part shown in Figure 7-30, but made with datum and feature holes to LMC size limits, or the largest-size holes, and a gage to check the upper control callout. This figure shows the effects of size and bonus tolerance on both the feature holes and datum feature H. As with the previous gage, effects of LMC and bonus tolerances (for the datum feature and feature holes) on the lower control callout would be similar, without the impact of datum H.

Continuing on, Figure 7-31b illustrates the effect of LMC holes and bonus tolerances impact on possible resultant thin walls. Always consider this possible result when applying tolerances at MMC. A simple formula is shown to aid in calculating this possible condition when doing a design analysis. A comparison of composite tolerancing and single segment tolerancing is shown later in the chapter.

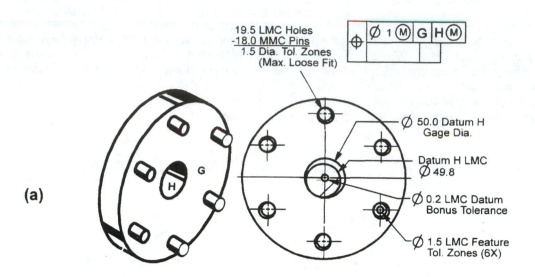

19.5 LMC Holes
−18.0 MMC Pins
1.5 Dia. Tol. Zones
(Max. Loose Fit)

⌖ | ⌀ 1 Ⓜ | G | H Ⓜ |

(a)

⌀ 50.0 Datum H
Gage Dia.

Datum H LMC
⌀ 49.8

⌀ 0.2 LMC Datum
Bonus Tolerance

⌀ 1.5 LMC Feature
Tol. Zones (6X)

Note: A functional gage accepts a 0.2 bonus tolerance at the
datum axis due to the datum feature being LMC, with a
0.5 bonus tolerance at each feature axis due to the
features being LMC, not a 0.7 tolerance at each feature
axis. Remember this principle when doing open setup
or CMM inspections. (See Chapter 8 and Figures
8-11 and 8-12.)

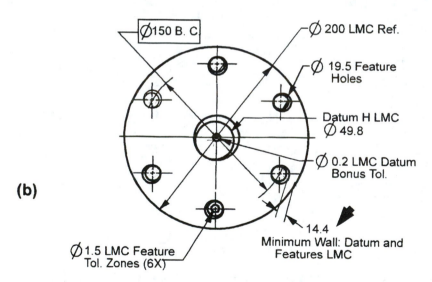

⌀150 B. C.

⌀ 200 LMC Ref.

⌀ 19.5 Feature
Holes

Datum H LMC
⌀ 49.8

⌀ 0.2 LMC Datum
Bonus Tol.

14.4
Minimum Wall: Datum and
Features LMC

(b)

⌀ 1.5 LMC Feature
Tol. Zones (6X)

Note: A minimum wall could evolve when features and datums are
LMC and may be determined as follows:

O.D. at LMC = 200 Ref.
    Less B.C. (150) Basic
         -   19.5 LMC Feature Holes
         -    1.5 Feature Hole Loc. Tol. LMC
         -    0.2 Datum Size Tol. LMC
         -    0   Datum Loc. or Orient. Tol. (if any)

$$\frac{28.8}{2} = 14.4 \text{ Min. Wall LMC}$$

**Figure 7-31   MMC Gage for Figure 7-30. (a) Part made LMC;
(b) impact on wall thickness.**

Figure 7-32 illustrates a possible gage when features are specified MMC with the datum hole reference RFS. Split rings or other variable gage features for the datum hole would be required. Although datum referencing at RFS is not impossible, it is very involved and can be quite expensive because variable gage elements would be required. Because of these reasons, RFS gaging is generally deemed impractical.

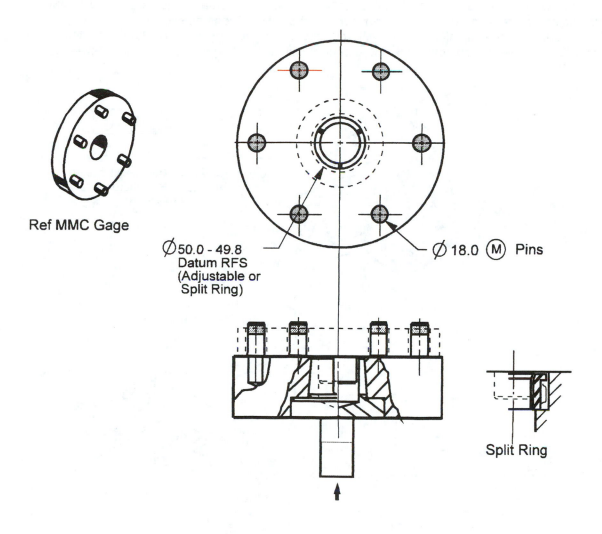

Ref MMC Gage

$\emptyset$50.0 - 49.8
Datum RFS
(Adjustable or
Split Ring)

$\emptyset$ 18.0 Ⓜ Pins

Split Ring

Figure 7-32   Possible gage features MMC; datum hole RFS.

Radial Hole Patterns: Primary Datum Axis It is possible to control a pattern of radially located holes relative to a primary datum axis by extending the composite position tolerancing principles. In Figure 7-33, the radial pattern of four 10 mm diameter holes is located from a primary datum axis A, a secondary datum surface B, and a tertiary datum slot C. The illustration depicts the four holes located from this framework within a 0.8 diameter tolerance zone when the feature holes are at MMC and the primary and tertiary datums are also at MMC. The tolerance zone is repeated four times at the basic dimensions as shown.

The lower callout is blank, so we will work through the design using a 0.2 diameter tolerance zone for the hole-to-hole tolerance (lower control callout), adding datum references as we go.

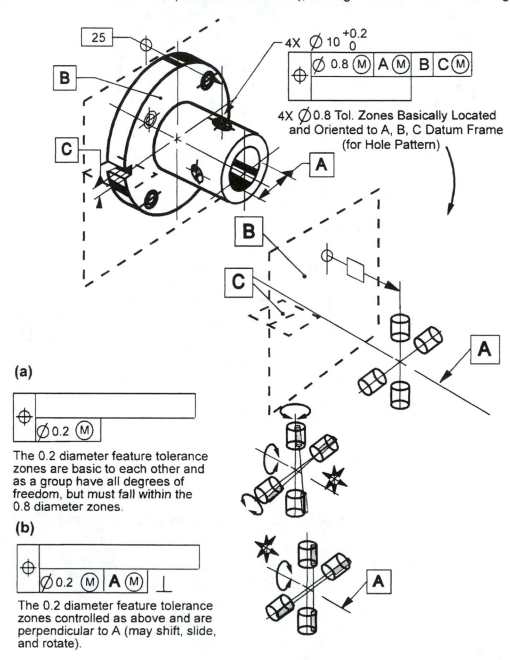

(a)

The 0.2 diameter feature tolerance zones are basic to each other and as a group have all degrees of freedom, but must fall within the 0.8 diameter zones.

(b)

The 0.2 diameter feature tolerance zones controlled as above and are perpendicular to A (may shift, slide, and rotate).

Figure 7-33   Radial hole pattern: primary datum axis

182

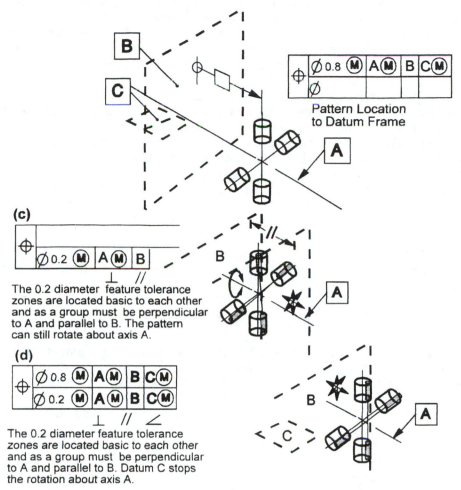

(c)

The 0.2 diameter feature tolerance zones are located basic to each other and as a group must be perpendicular to A and parallel to B. The pattern can still rotate about axis A.

(d)

The 0.2 diameter feature tolerance zones are located basic to each other and as a group must be perpendicular to A and parallel to B. Datum C stops the rotation about axis A.

Conclusion:
The upper control frame serves to establish the boundaries for the location tolerance control of the hole pattern.

The lower feature control frame serves to establish the tolerance zones for the individual feature-to-feature relationship, and this relationship to the upper control frame.

Figure 7-33 (continued)

In Figure 7-33a, there are no datum references specified. Therefore, the lower control tolerance zones of the pattern of four holes (0.2 MMC) is totally free to float, slide, twist, and rotate, as long as they remain within the 0.8 diameter tolerance zones of the upper control.

In Figure 7-33b, datum A has been invoked, and because datum feature A is a feature of size, it is subject to size variation. Therefore MMC has been specified to clarify how datum A references apply. Per Figure 7-33b, the 0.2 diameter tolerance zones must be *oriented* to datum axis A; that means that the only orientation possible to axis A is *perpendicularity*. The feature pattern 0.2 tolerance zones are free to slide and rotate, within the 0.8 diameter zones, but must remain perpendicular to A.

In Figure 7-33c, secondary datum surface C has been added, which requires that the axis of the four 0.2 diameter tolerance zones be *oriented*, in this case parallel, to surface B. We recall that datums are defined as mutually perpendicular, but datum features will have processing error and are not perfect; therefore, some out of perpendicularity between datum features A, B, and C will exist.

This error will naturally get reflected into the secondary datum surface B. By controlling the holes to primary datum axis A, we achieve a degree of automatic parallelism to within the allowable perpendicularity limits imposed between datums A and B. There exists some question about the value here of datum reference B, which could be argued as redundant, but the importance of specifying the precise *primary datum* cannot be overemphasized.

Figure 7-33d has added tertiary slot C to the control callout, which stops all rotation of the four-hole pattern. The completed lower control callout states that the 0.2 diameter tolerance zones must be perpendicular to datum A MMC, parallel to datum surface B, with any remaining rotation stopped by datum C MMC. This completed callout is now identical in content, top and bottom. This callout may appear redundant and probably is, using the former interpretations from ANSI Y14.5M-1982. Because the ASME Y14.5M-1994 standard states the lower control is orientation (not pattern location), having both callouts identical in this case is not redundant, just different. These circumstances are unique to cylindrical parts with the primary datum as a feature axis, with the orientation definition applied. This exercise illustrates the importance of thoroughly analyzing the design, with a working knowledge of the current ASME standard, and the *composite positional tolerancing* process before applying controls and datum frameworks to unique parts or assemblies. If a datum is not needed, do not invoke it! A comparison of composite and single segment position controls for radial hole patterns is illustrated in Figure 7-37 and 7-38 later in the chapter.

Composite versus Multiple Single Segment Position Controls

To help clarify these issues, let's compare the two concepts. A *single segment position control* establishes the *location and orientaion* of feature relationships to datum frameworks. The use of multiple controls does not change this concept; we simply add or omit datums for associated controls.

Composite positional controls per the 1994 standard invoke the orientation only principle in the lower control frame and use only one position symbol for two separate purposes. The upper control is the same as a single segment control in that it establishes the pattern location and orientation from the datum framework, whereas the lower control establishes the feature relationship and the orientation (not location) of this relationship to the datum(s) specified. These are distinct concepts.

The following figures will be used for comparisons of these concepts:
   Rectangular pattern: Figure 7-7a (metric)
   Aligned Pattern: Figure 7-27
   Circular Pattern: Figure 7-29
   Radial Pattern: Figure 7-33

In Figures 7-34 through 7-38, all controls are MMC for the features and RFS for the datums. There could be bonus location tolerances for the features or datums based on their size, but let us concentrate on a single concept for this exercise. Compare the concepts, controls, datums, and symbols used and note the results of both concepts.

Multiple Single Segment Controls

If different datum references are specified or if datums in a differing order of precedence are required, *single segment position controls* are to be specified, which is done with two separate control frames. This technique constitutes completely independent design requirements. The position symbol is entered twice as shown in Figure 7-39. As was shown in the previous exercise, the position controls are independently verifiable, and both the upper control and lower control deal with *location and orientation*. The composite position symbol is not to be applied.

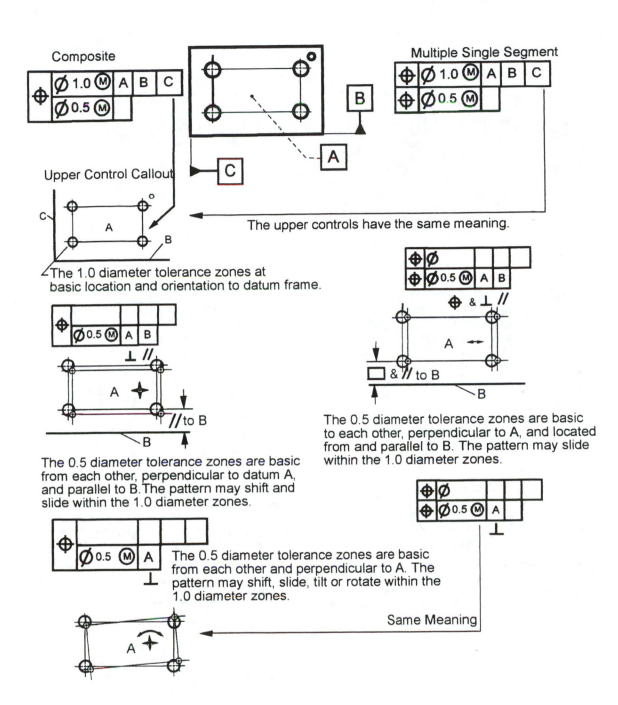

Composite

The 1.0 diameter tolerance zones at basic location and orientation to datum frame.

Upper Control Callout

Multiple Single Segment

The upper controls have the same meaning.

The 0.5 diameter tolerance zones are basic from each other, perpendicular to datum A, and parallel to B. The pattern may shift and slide within the 1.0 diameter zones.

The 0.5 diameter tolerance zones are basic to each other, perpendicular to A, and located from and parallel to B. The pattern may slide within the 1.0 diameter zones.

The 0.5 diameter tolerance zones are basic from each other and perpendicular to A. The pattern may shift, slide, tilt or rotate within the 1.0 diameter zones.

Same Meaning

Figure 7-34 Ref. Figure 7-7a (metric). Analysis: composite versus multiple single segment position tolerancing.

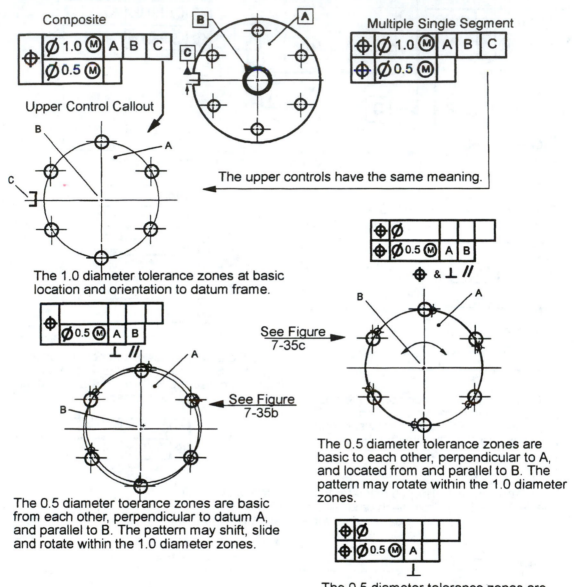

Composite

⊕	Ø 1.0 Ⓜ	A	B	C
	Ø 0.5 Ⓜ			

Upper Control Callout

Multiple Single Segment

⊕	Ø 1.0 Ⓜ	A	B	C
⊕	Ø 0.5 Ⓜ			

The upper controls have the same meaning.

The 1.0 diameter tolerance zones at basic location and orientation to datum frame.

⊕	Ø			
⊕	Ø 0.5 Ⓜ	A	B	

⊕ & ⊥ //

⊕				
	Ø 0.5 Ⓜ	A	B	

⊥ //

See Figure 7-35c

See Figure 7-35b

The 0.5 diameter tolerance zones are basic from each other, perpendicular to datum A, and parallel to B. The pattern may shift, slide and rotate within the 1.0 diameter zones.

The 0.5 diameter tolerance zones are basic to each other, perpendicular to A, and located from and parallel to B. The pattern may rotate within the 1.0 diameter zones.

⊕	Ø			
⊕	Ø 0.5 Ⓜ	A		

⊥

The 0.5 diameter tolerance zones are located from and oriented to surface A only; therefore, perpendicularity is the only control possible.

(a)

**Figure 7-35   See Figure 7-29.
Analysis: composite versus multiple
single pegment Position tolerancing.**

186

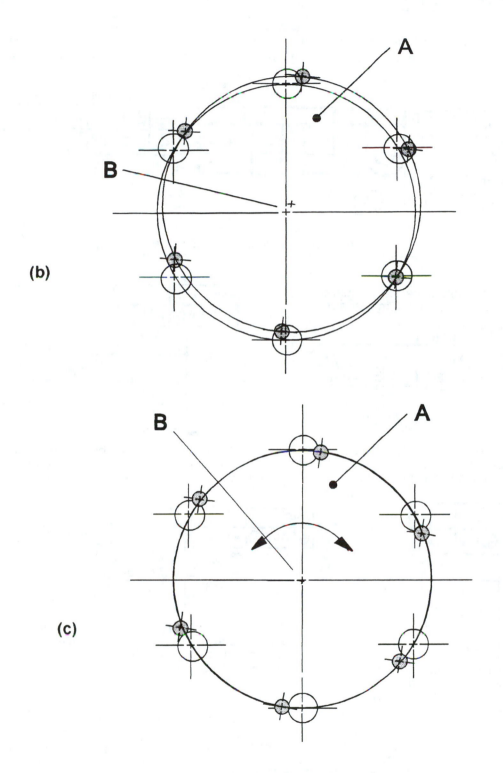

(b)

(c)

Figure 7-35 (continued)

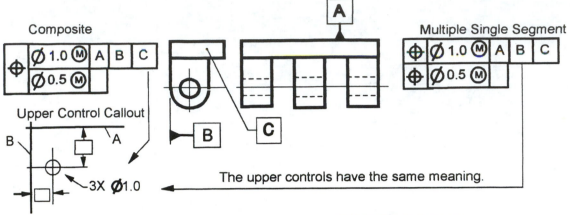

Composite

Multiple Single Segment

Upper Control Callout

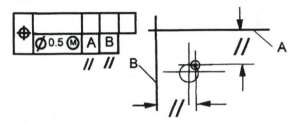

The upper controls have the same meaning.

The 1.0 diameter tolerance zones must be aligned and located from datum framework.

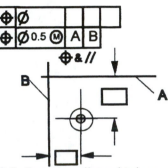

The 0.5 diameter aligned tolerance zones must be oriented (parallel) to datums A and B and be within the 1.0 diameter zones.

The 0.5 diameter aligned tolerance zones must be both located and oriented to the datum frame, which results in a bull's-eye or redundant callout.

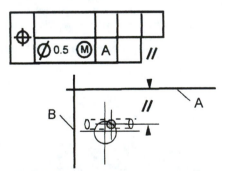

The 0.5 aligned zones may slide and twist within the 1.0 dia zones and be parallel to datum A only.

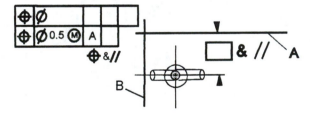

The 0.5 diameter aligned tolerance zones may slide and twist within the 1.0 diameter zones, but must be located from and oriented to A.

Figure 7-36 See Figure 7-27.
Analysis: composite versus multiple single segment position tolerancing.

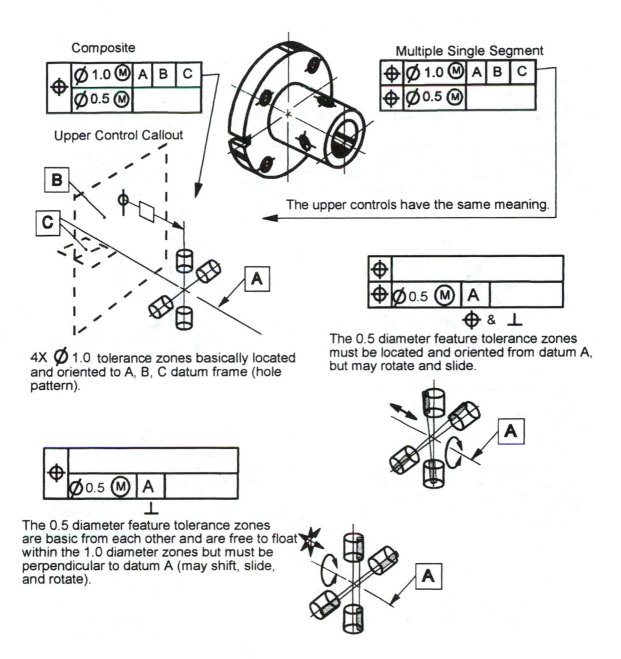

Composite

⊕	⌀ 1.0 Ⓜ	A	B	C
	⌀ 0.5 Ⓜ			

Upper Control Callout

Multiple Single Segment

⊕	⌀ 1.0 Ⓜ	A	B	C
⊕	⌀ 0.5 Ⓜ			

The upper controls have the same meaning.

B

C

A

4X ⌀ 1.0 tolerance zones basically located and oriented to A, B, C datum frame (hole pattern).

⊕			
⊕	⌀ 0.5 Ⓜ	A	

⊕ & ⊥

The 0.5 diameter feature tolerance zones must be located and oriented from datum A, but may rotate and slide.

A

⊕			
	⌀ 0.5 Ⓜ	A	

⊥

The 0.5 diameter feature tolerance zones are basic from each other and are free to float within the 1.0 diameter zones but must be perpendicular to datum A (may shift, slide, and rotate).

A

Figure 7-37 See Figure 7-33.
Analysis: composite versus multiple single segment position tolerancing.

Composite

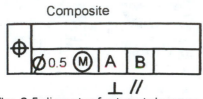

The 0.5 diameter feature tolerance zones are basic from each other, and are free to float within the 1.0 diameter zones but must be perpendicular to datum A and parallel to datum B.

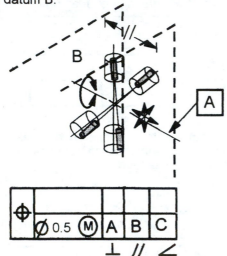

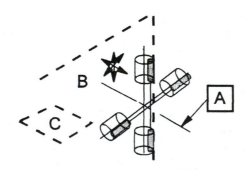

The 0.5 diameter feature tolerance zones are located basic to each other, and as a group must be perpendicular to A and parallel to B. Datum C stops the rotation about axis A.

Multiple Single Segment

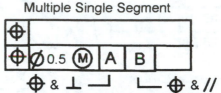

The 0.5 diameter feature tolerance zones are located basic to each other and as a group must be located from and perpendicular to A and located from and parallel to B. The pattern can still rotate about axis A.

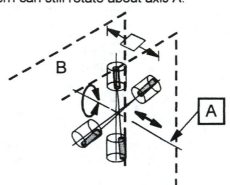

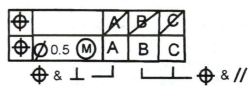

Redundant control: The 0.5 diameter feature tolerance zones must be located from and oriented to the datum frame in both callouts, a bull's-eye.

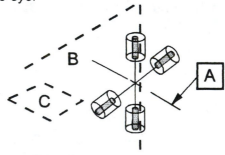

Figure 7-38 See Figure 7-33.

Analysis: composite versus multiple single segment position tolerancing.

Least Material Condition (LMC)

Positional tolerance controls may be applied for features at LMC. For holes, this would be the largest possible hole within the size limits. The most common application for this control is to control minimum wall thickness. Controls specified at MMC allows a bonus location tolerance as holes get larger. This bonus tolerance, along with increased hole size, may result in thin walls at the edges of a part or when holes are centered about or within bore or shaft diameters. With LMC, the control tolerance is applied to the largest hole size, with bonus tolerance applying as the hole gets smaller. For shaft diameters, LMC may be applied, but the size conditions and bonus application is reversed. Bonus tolerance gets added as the feature departs from LMC size as shown in Figure 7-40.

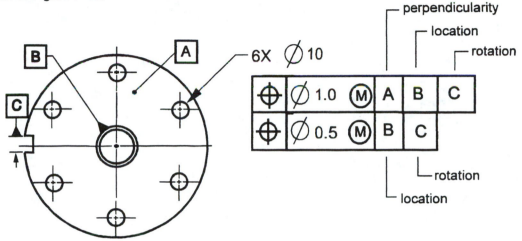

Figure 7-39 Multiple single segment controls.

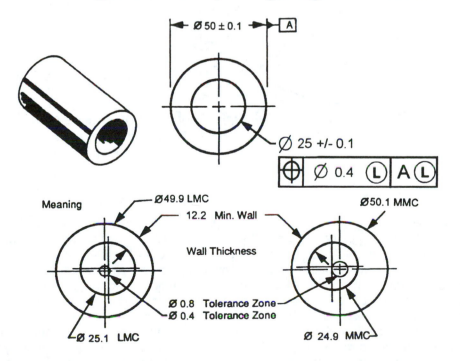

Figure 7-40 LMC position controls: effects on wall thickness.

191

CONCENTRICITY

Concentricity is the condition where the *median points* of all diametrically opposed elements of a feature surface of revolution (cylinders, hexagons, cones, spheres, cubes, etc.) are coaxial with a datum feature axis. A *concentricity* tolerance zone is a cylindrical or spherical tolerance zone whose axis (or centerpoint) coincides with the axis (or centerpoint) of the datum feature. The median points of all corresponding elements of the feature(s) are controlled, RFS, and must lie within the cylindrical (or spherical) zone. Concentricity is considered a location control in that it controls a derived feature axis relative to a datum axis. The ASME standard uses the more formal term "derived median line" relative to a datum axis. The median line must be derived from surface elements and is always RFS. Concentricity is considered difficult to measure due to the description and derivation of the tolerance zone. See Figure 7-4.

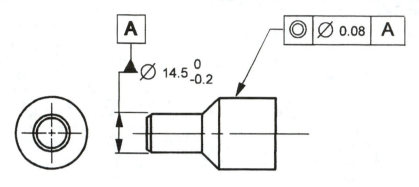

Meaning

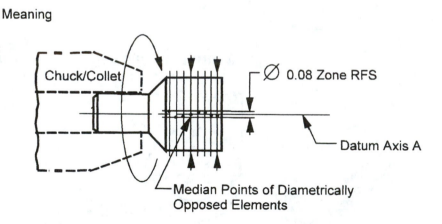

Figure 7-41   Concentricity.

Because the tolerance zone is at the feature (median line) axis, it is required to derive the line or axis, working through the feature surface. Figure 7-42a illustrates the use of a single indicator device as an example for obtaining measurements. With a single indicator, it can be difficult to distinguish between form (circularity) error and location error. Without further assessment, it is unclear if the feature is out of location or not round. Further, a feature could be within size limits but not be within the circularity limits and be perfectly concentric, yet be rejected for possibly the wrong reasons.

A common method for evaluation of concentricity is with two indicators in opposed locations. Using the methods and formula of Figure 7-42b, a locus of points may be plotted for any degree increments in one full revolution of the feature about the datum axis. X is the datum axis origin in the setup and may be of any known value. As the measuring device moves along the feature,

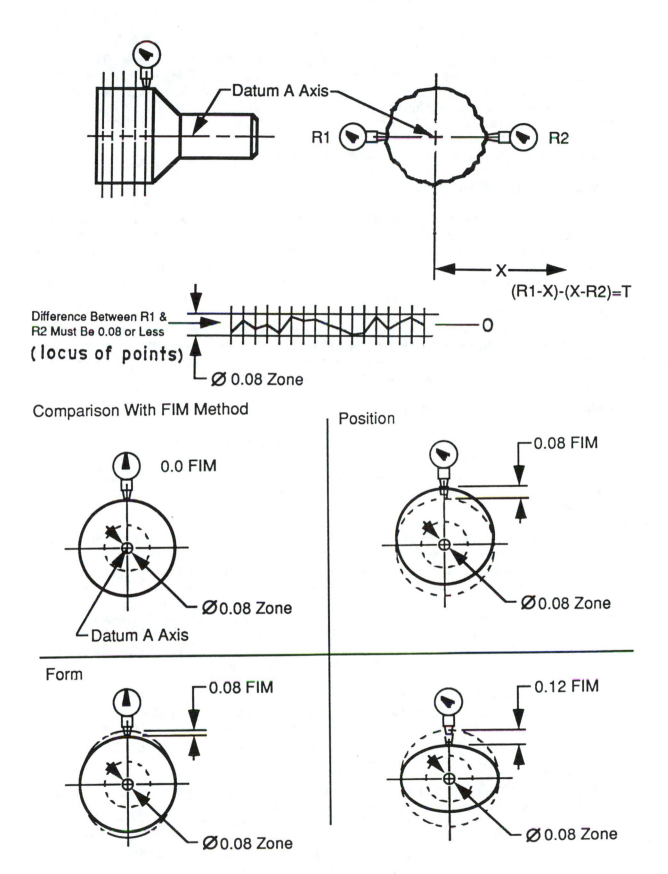

Figure 7-42 Concentricity evaluation.

193

the locus of points gathered at each measured section of the surface constitutes the derived median line and must fall within the tolerance zone cylinder specified. This method works well except for parts with features of an odd number of lobes. Lobbing is found in parts manufactured by some grinding methods, and lobs may generally number anywhere from 3 to 19. A computerized program is helpful in these cases. More advanced techniques will also work, including air/electronic gaging, computer assist measurement, and high-resolution magnification. These methods generally are more complicated and costly than other methods.

Concentricity and Datum Error

By definition, concentricity is relative to a *datum axis*. This means that the datum feature may have size variation. If *vee locators* are used to isolate the datum feature, we must be aware of the setup and the potential impact on measurement. Datum errors could be read at the surface being measured, possibly affecting acceptance/rejection decisions. The same conditions that affected runout readings may also impact *concentricity* if evaluated in the same manner. Figures 7-43 and 7-44 are reminders of the results of datum error, vee locator type, and indicator(s) location. As with runout, the inspection setup is critical in order to achieve repeatable measurement results.

Because of the complexity of finding the derived median line of a feature, and with all the influential variables involved, it is understandable why position or runout might be preferred over concentricity in a production or factory environment. By its definition, concentricity is best suited to a lab environment or where sophisticated equipment is in place.

Selection of Coaxial Controls

Coaxiality is that condition where axes of two or more surfaces of revolution are coincident. The amount of permissible variation from coaxial may be expressed by a *position, runout, profile, or concentricity control.* Generally, a design configuration along with some simple logic will help determine which coaxial control to use. Figure 7-45 illustrates some key reminders to help make that determination.

Runout: A *composite error* read at a *feature surface*, RFS, acceptable when other controls of size and/or form or orientation are in place. Circular and total runout controls are available; however, total runout requires the ability in the setup to traverse the indicator device parallel or perpendicular to the datum axis.

Profile: Controls a *feature surface* size, form, orientation, and location to a datum axis. Profile is applied to the feature RFS, but may be applied to the datum RFS or MMC. Both *profile of a line* and *surface* controls are available. Profile may also be applied without datum references, a very versatile control.

Position: Normally applies to feature axes or centerplanes, and may be applied MMC, LMC, or RFS. MMC and LMC will allow bonus tolerances. Applied MMC, position also provides ability to develop functional gages, if desired. Position is generally applied to static fit assemblies and hole patterns. It creates a virtual mating envelop when applied MMC.

Concentricity: As it applies to a feature derived median line relative to a datum axis, it is not as commonly used as it once was. In the past concentricity was often abused and misapplied as "concentric TIR" (runout). For reasons previously mentioned, position, runout, or profile are recommended over concentricity. Typical past applications of concentricity have been to control the axis of a rotational mass to a datum axis.

194

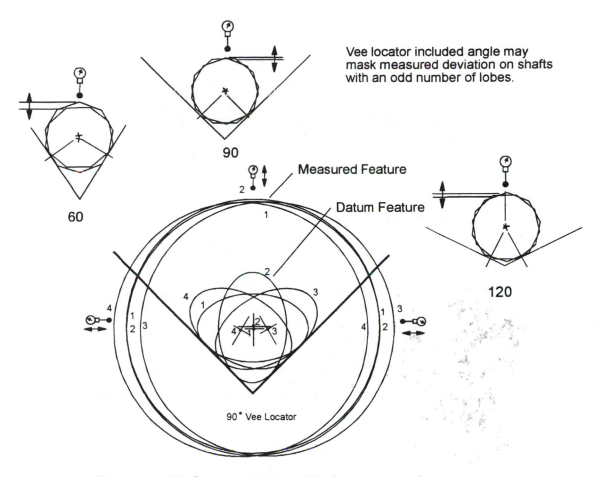

Vee locator included angle may
mask measured deviation on shafts
with an odd number of lobes.

60

90

120

Measured Feature

Datum Feature

90° Vee Locator

Figure 7-43 Concentricity: 90 degree vee locators.

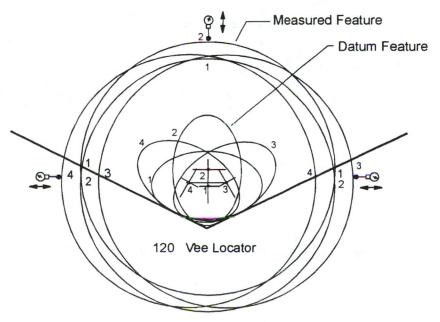

Measured Feature

Datum Feature

120   Vee Locator

Figure 7-44 Concentricity: 120 degree vee locators.

Coaxial Controls

Runout ✓ ✓✓

    Allows composite surface error (size, form, location)

    Always RFS

    Always datum related

Position ⊕

    Applied MMC, RFS, LMC; tolerance zone at feature axis

    Functional gaging allowed when applied MMC (mating envelope)

    MMC & LMC allow bonus tolerances

Profile (Line or Surface) ⌒

    May control Size, Form, Orientation, and Location

    Applied RFS

    Applied with or without datum references

Concentricity ◎

    For control of feature *median points* only, related to datum axis (involves a derived median line)

    Generally associated with dynamic balance

    Often misapplied

Figure 7-45 Selection of coaxial controls.

196

SYMMETRY

Symmetry is that condition where all median points of all opposed elements of a feature surface, or corresponding elements of two or more feature surfaces are congruent with the axis or centerplane of a datum feature. Symmetry controls were removed from the 1982 Y14.5 standard because it was felt that symmetry could be achieved by the use of true position RFS. With the updated definitions we now have, this is no longer the case. Symmetry and concentricity are similar concepts, and deal with the location of median lines or points. The concepts do not deal with maximum inscribed, or minimum circumscribed cylinders, which are the foundation of position controls.

The omission of symmetry left the U.S. and ISO standards not in sync, because ISO continued the use of the symmetry control. Many felt that with the addition of ASME Y14.5.1M standard, and because of the more precise definitions in place, it was necessary to return symmetry as a control for feature median planes. See Figure 7-46. With the return of symmetry, the U.S. and ISO standards are more in line, yet differences still exist in the exact wording of its definition.

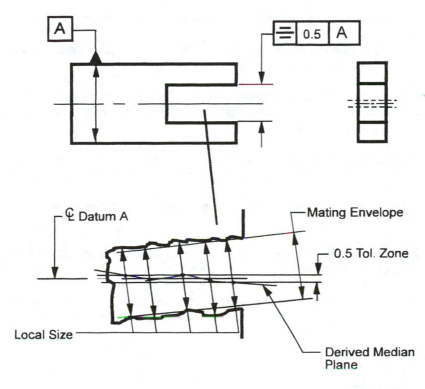

Within the limits of size, RFS, all median points of opposed elements of the feature slot must lie between two parallel planes 0.5 apart, the two planes being equally disposed about datum plane A.

Figure 7-46   Symmetry RFS.

LOCATION CONTROLS SUMMARY

All coaxial or symmetrical features shall be controlled by position, runout, profile, and concentricity/symmetry. Consider runout or position first.

Concentricity tolerance requires establishment of a feature's "derived median points".

Where a cylindrical tolerance zone is intended, a diameter symbol is to be used.

Where a spherical tolerance zone is intended, the spherical radius or diameter symbols (SR or SO) are to be used.

Bidirectional position tolerancing is used to express a rectangular tolerance zone.

Consider the use of the *projected tolerance zone* on fixed fastener applications where mating part thickness is greater than threaded or other secured feature depth.

Zero tolerancing offers maximum manufacturing flexibility, but it complicates the variables' data measurement process. The use of zero tolerancing is recommended only when manufacturing and quality areas are directly involved.

Composite positional tolerancing invokes *orientation* controls in the lower control frame. Multiple single segment position controls invoke *location* in both the upper and lower control frames.

Bonus tolerances may have an adverse effect on wall thickness; therefore, consider the use of LMC controls.

For features of size, consider the impact of virtual/resultant conditions, that is, the collective extremes of size and geometric tolerances.

When using *composite tolerancing* while applying a *projected tolerance zone*, the projected tolerance zone symbol is applied to the lower frame, following the material condition symbol, if any.

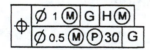

EXERCISE 7-1 Position Tolerance

True or False, or fill in.

1. True Position tolerance controls are implied MMC, unless otherwise specified. T F

2. A True Position tolerance zone is a +/ - zone. T   F

3. (M) allows a bonus location tolerance.   T   F

4. For hole pattern locations, datum references are required. T   F

5. Indicate the correct positional tolerance value associated with each size tolerance for the figure below.

6. The MMC virtual condition for the holes below is _____ .

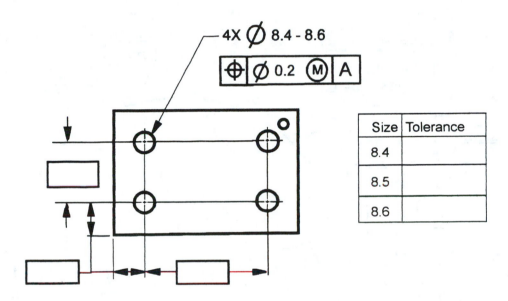

Size	Tolerance
8.4	
8.5	
8.6	

Complete the formulas for finding:

Floating fastener positional tolerance:

Fixed fastener positional tolerance:

Hole diameter at MMC virtual ondition:

Shaft diameter at MMC virtual condition:

EXERCISE 7-2 Positional Tolerance Application

The design shown in this exercise is a portion of a crankshaft, with a timing gear assembled. The gear is a close free fit on the shaft, and a free fit over a key that correctly locates the gear timing tooth. The assembly is secured by four 10 mm mounting bolts. With the information from the exercise on page 199, complete the following:

(a) Determine the feature controls/tolerances for this design.
Apply 60% of the available tolerance to the shaft/key assembly and the remaining 40% to the gear. Apply all tolerances to both features and datums at MMC. Also determine the maximum hole size allowed for the gear.

(b) Determine the information required for the receiver gage designs shown on pages 201 and 202. In calculating gage features size, ignore gaging tolerances for now.

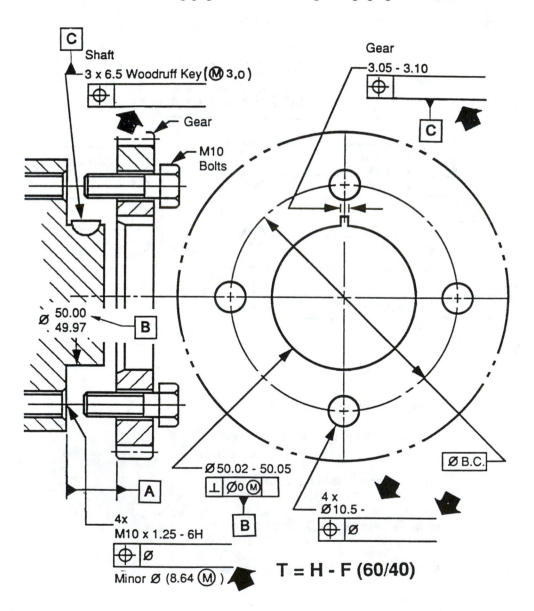

GEAR GAGE

(c) Determine the key width, shaft bore diameter, and 4 gage pin diameters.

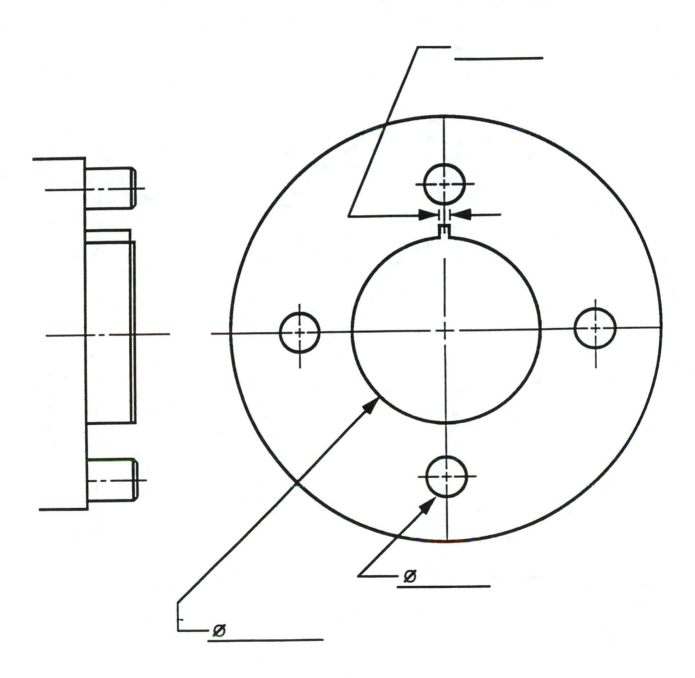

Ø

Ø

EXERCISE 7-2 (Continued)

SHAFT GAGE

(d) Determine the keyway width, shaft nose diameter, and 4 gage pin diameters for the shaft gage.

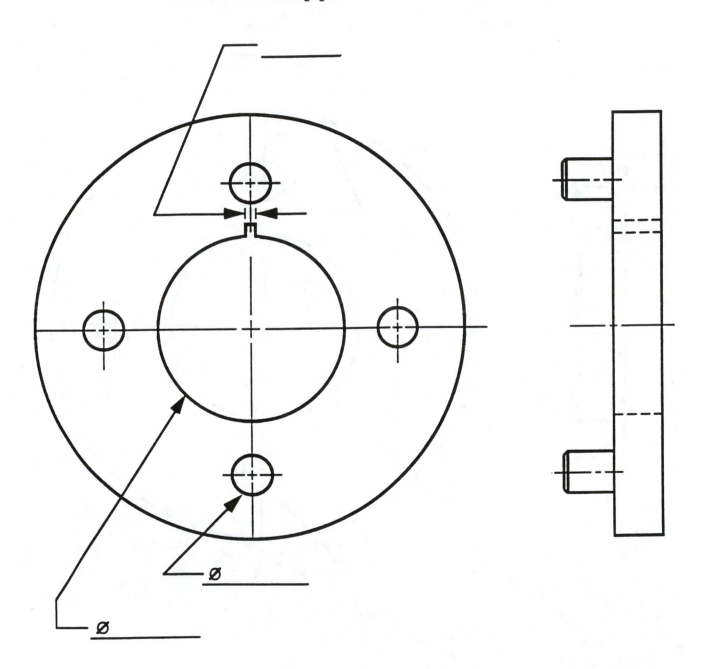

202

EXERCISE 7-3 Position Applications

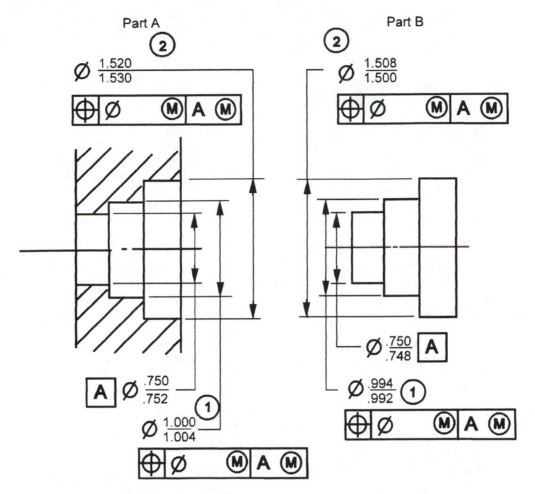

Determine tolerance for bore and pin diameters #1. Apply equally.

Determine tolerance for bore and pin diameters #2. Apply equally.

What are MMC virtual conditions for each set of features?

Part A diameter 1_____   Part B diameter 1_____

Part A diameter 2_____   Part B diameter 2_____

EXERCISE 7-3   (continued)

When using *zero tolerancing* principles, the clearance between mating diameters at MMC must be equal to or greater than the size tolerance difference of both mating datum features at MMC. That is, .752 - .750 = .002. Therefore,

1.508 + .002 = 1.510 MMC

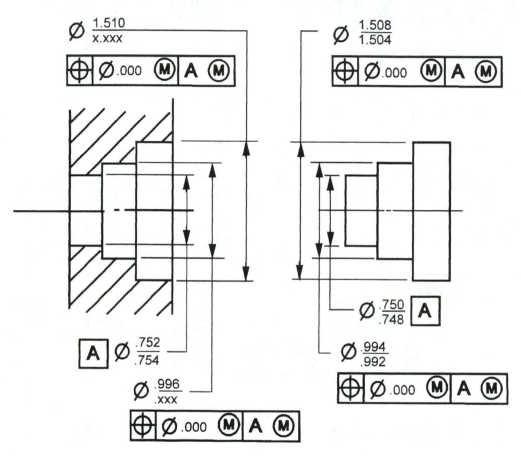

.994 + .002 = .996 MMC

EXERCISE 7-4 Location of hole patterns.

The following figures represent a similar part with a center hole along with a pattern of four small holes. Follow the directions for each figure.

Complete the feature control callouts.

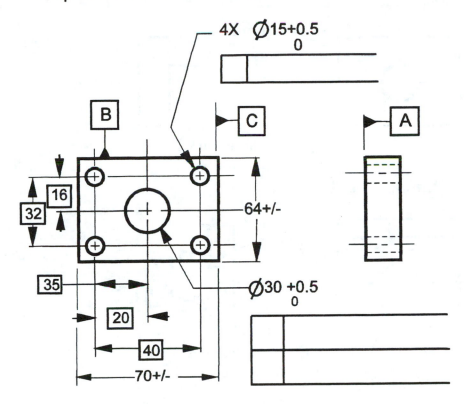

Functional gaging is to be used for these requirements.

The large center hole is to be located from datum framework A, B, C within a diameter tolerance of 0.8. It is also to be perpendicular to datum surface A within a diameter tolerance of 0.5.

The large center hole is to be datum D.

The four small holes at MMC are to be relative to datum A and datum D MMC within a diameter tolerance of 0.5.

EXERCISE 7-5 Gages for Ex.7-4

Determine the gage pin diameters in the figures below.

Perpendicularity Gage

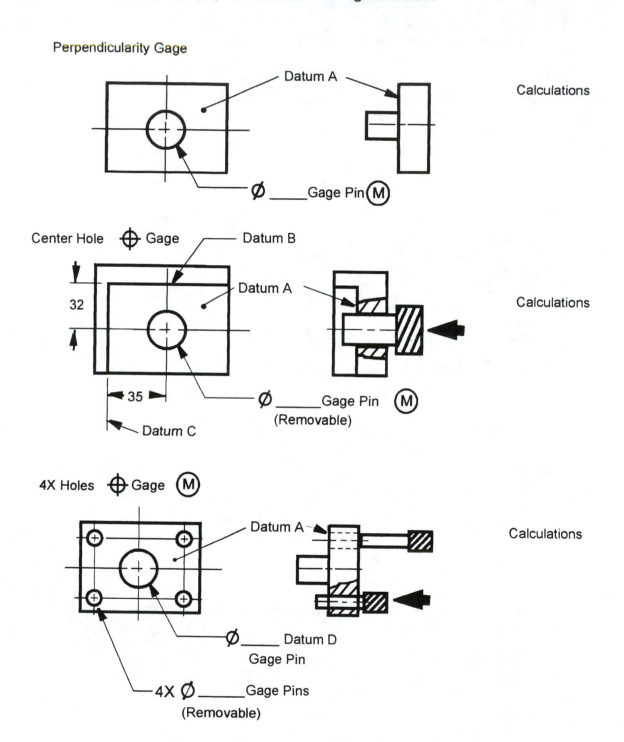

Datum A

Ø _____ Gage Pin (M)

Calculations

Center Hole ⊕ Gage

Datum B

Datum A

32

35

Datum C

Ø _____ Gage Pin (M)
(Removable)

Calculations

4X Holes ⊕ Gage (M)

Datum A

Ø _____ Datum D
Gage Pin

4X Ø _____ Gage Pins
(Removable)

Calculations

EXERCISE 7-6 Altered requirements for Ex. 7-4

Adjust the control callout so that the four small holes are MMC
but relative to surface A, the large hole RFS, and tertiary datum
surface B.

Tolerance zones are similar to Exercise 7-5.

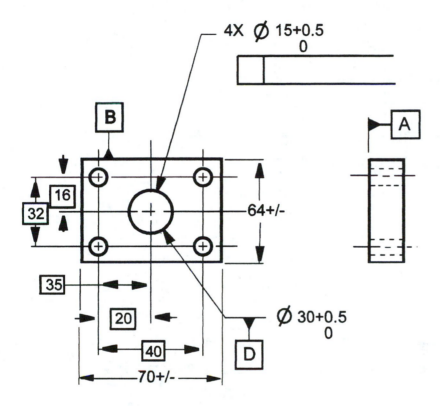

4X Ø15+0.5
0

| ⊕ | Ø 0.5 Ⓜ | A | D | B |

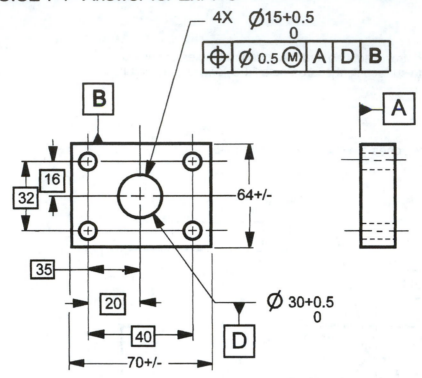

B

A

16

32

64+/-

35

20

40

70+/-

Ø 30+0.5
0

D

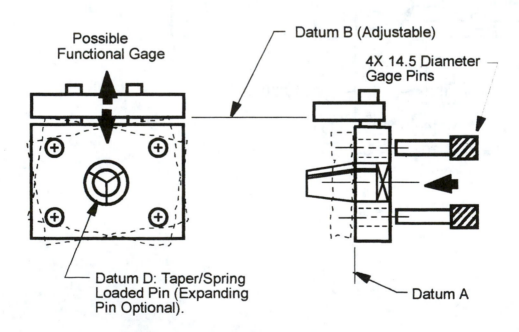

Possible
Functional Gage

Datum B (Adjustable)

4X 14.5 Diameter
Gage Pins

Datum D: Taper/Spring
Loaded Pin (Expanding
Pin Optional).

Datum A

Note that the tertiary datum surface of the gage is
adjustable when the secondary datum feature
has an axis. (The tertiary datum can only stop
the rotation of the workpiece in the gage fixture.)

EXERCISE 7-8   Gaging RFS

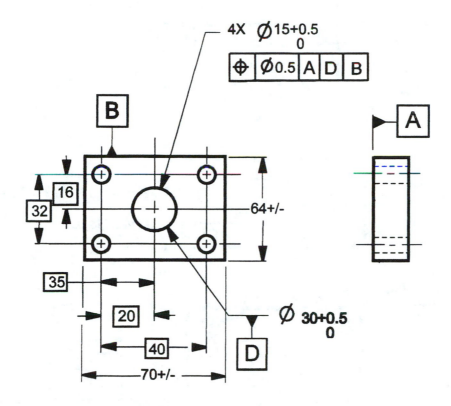

4X ⌀15+0.5 / 0

⊕	⌀0.5	A	D	B

B

16

32

64+/-

35

20

40

70+/-

D

A

⌀ 30+0.5 / 0

The principles of functional gaging cannot be utilized when both the feature and datum are specified RFS. Open setup inspection, CMM, or the techniques of paper gaging would be necessary.

EXERCISE 7-9 LMC Gaging

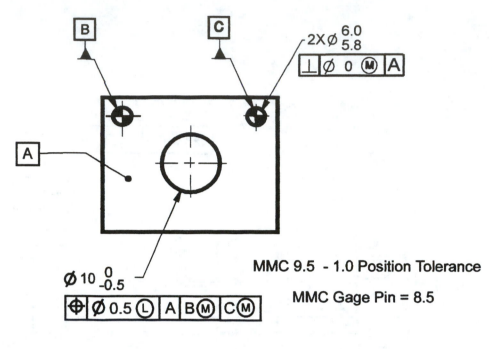

MMC 9.5 - 1.0 Position Tolerance

MMC Gage Pin = 8.5

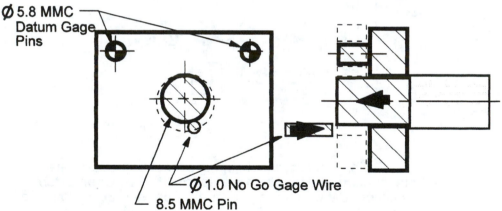

The center gage pin is calculated by developing the positional tolerance at MMC (1.00) and subtracting the value from the features MMC size limit.

Thus a 8.5 diameter MMC gage pin is needed for the large center hole. A further gage wire that is equivalent to the MMC tolerance zone of 1.0 diameter is required.

When the part is placed on the MMC gage, located on datum pins B and C, the large 8.5 center pin must GO in the part, but the 1.0 NO GO gage wire must never go between pin and hole.

See MIL-HDBK 204A(AR) for further info.

EXERCISE 7-10 Concentricity

True or False, or fill in.

1. Concentricity is always datum related.   T   F

2. Concentricity is applied _____ . (MMC, LMC, RFS)

3. The tolerance zone exists at the feature _____ and is related
   to the datum _____.

4. The use of vee locators, for datum set-up is recommended.   T   F

5. Because concentricity controls require the verification of feature axes
   to datum axes, without regard to surface conditions, the use of runout,
   position RFS, or profile controls would be a better choice in most
   cases. T   F

6. In the figure below, illustrate the two larger diameters to be Concentric
   with the smaller diameter within 0.1 mm. Draw the tolerance zones.

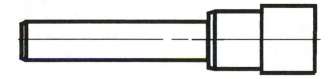

8 PAPER GAGING

Paper gaging is merely the graphical and mathematical manipulation of inspection data, derived by a measurement means other than functional or fixed dedicated gaging. Once we have used the formulas for tolerancing and know how to develop simple gages, we need to be able to use the same concepts to accept or rejects parts of the same quality as a functional gage, but measured from some other methods, such as open setup surface plate inspection or coordinate measuring machines.

Time constraints, volume, or gage expense may prohibit the use of dedicated gaging. The process control system development may require audits for feedback to the engineering or manufacturing areas. For these and other reasons, it is necessary to understand the principles of *paper gaging analysis*. In the examples that follow, we will walk through a typical inspection process using paper gaging principles.

MULTIPLE HOLE PATTERNS

Before we start, let's look at a coordinately dimensioned (+/-) part with a six-hole pattern of holes. Figure 8-1 illustrates a hole pattern located from the part edges by two 1.000 +/-.015 dimensions. The hole to hole tolerance is 2.000 +/- .010 and 4.000 +/- .010.

To evaluate this part without a receiver gage, it will be necessary to find out if the pattern of six holes falls within the pattern location tolerance of 1.000 +/- .015 from the part edge in both directions (see Figure 8-1b). The most common method is to isolate on the part edge to see if the holes fall within the 1.000 +/- .015 tolerance (both vertical and horizontal), as shown in Figure 8-1b. Next, evaluate the hole-to-hole dimensions/tolerance of 2.000 +/- .010 and 4.000+/- .010 by locating in one of the corner holes of the pattern, and by locating in second hole some distance away to establish a horizontal line 0 - 0. From this horizontal line may be established a second vertical line (perpendicular to the first), thus giving us the X and Y reference lines. From this setup, we may take measurements to find actual hole locations of remaining holes of the pattern as shown in Figure 8-1b.

Figure 8-2 illustrates the holes plotted on graph paper of 1/4 inch squares, with a scale of 1/4 in = .002 (or 125:1 visual ratio). The center of the plot in X and Y is also the theoretically perfect center of each individual hole. The location dimensions have been collapsed so that each hole is theoretically resting on top of the others. All holes lie within the +/- .015 pattern tolerance, and the part is acceptable so far.

Does each hole fall within the +/- .010 feature tolerance, and is the part acceptable? Will the part function as intended? Using coordinate methods of dimensioning and giving one hole a tolerance of zero, we might conclude that hole 6 is out of location tolerance and therefore the part is not acceptable. Using this method of dimensioning and analysis without further evaluation of the data will most likely result in the rejection of possibly good parts.

In the inspection process, we may not always pick the same holes to be zero, or 1. Different inspectors may use 1 and 3, 3 and 1, 4 and 6, or 6 and 4. By making one hole zero, we remove the +/- .010 tolerance allowed for that hole, and even though *any hole* could be zero, the coordinate dimensioning method has no provision for allowing this tolerance. Because we have no control over which hole the inspector selects, we must have some way of allowing the tolerance for holes 1 and 3.

One technique is the use of a transparent overlay made to the same scale as the plot (1/4 = .002). By adjusting the .010 overlay tolerance zone .002 left and .004 vertical, we establish a location that will accept all hole locations, and still remain within the +/- .015 pattern location tolerance. This technique allows the tolerance for hole 1 to be returned, provided that the pattern location tolerance is not violated. This concept is illustrated in Figure 8-3.

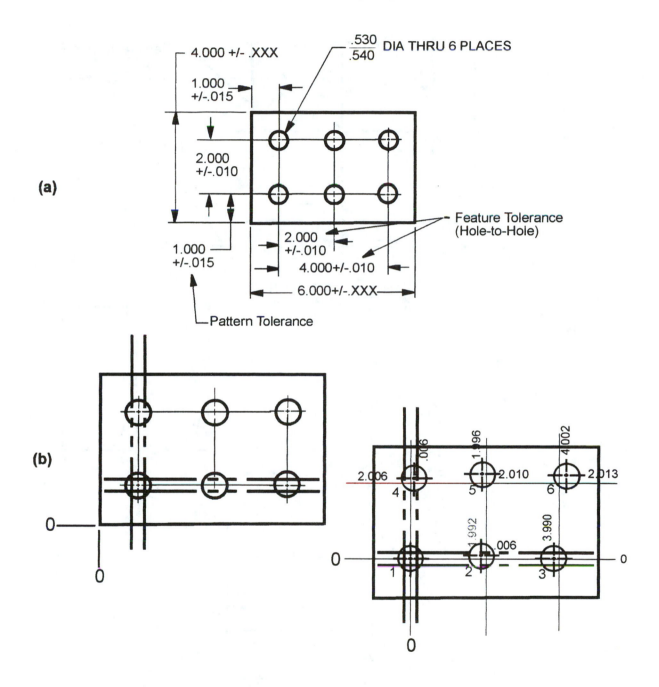

Figure 8-1 Paper gaging techniques: open setup inspection.

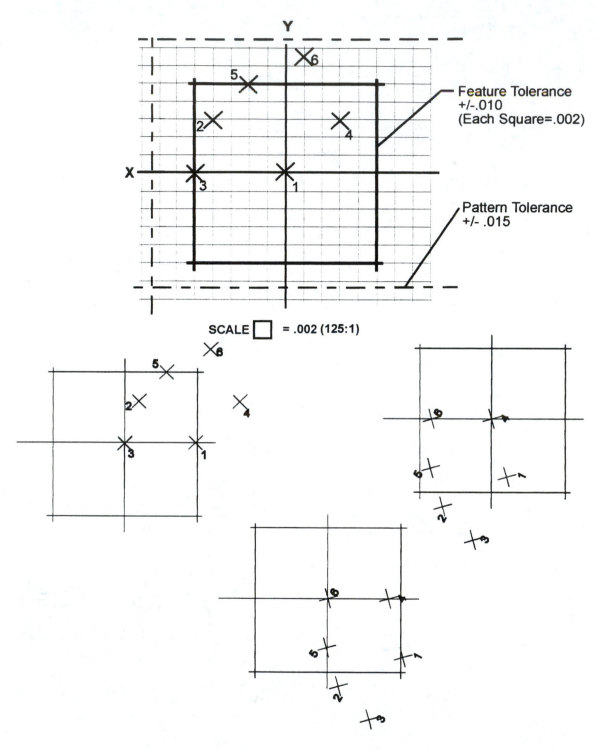

Feature Tolerance
+/-.010
(Each Square=.002)

Pattern Tolerance
+/- .015

SCALE ☐ = .002 (125:1)

Figure 8-2   Paper gage plot: six-hole pattern.

214

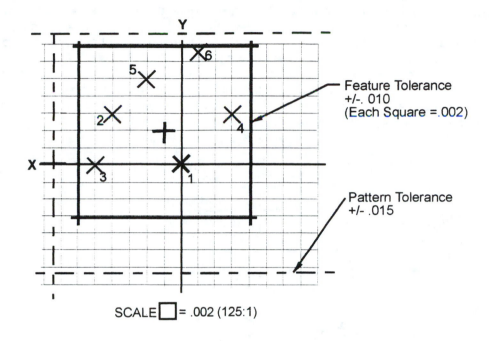

SCALE ☐ = .002 (125:1)

Figure 8-3   Paper gage plot: adjusted best-fit pattern location.

Evaluate Figure 8-3 and ask:

Has the pattern tolerance of +/- .015 been met? Yes!
Has the hole-to-hole tolerance of +/- .010 been met? Yes!
Will it function as intended? Yes!

The part illustrated by Figure 8-3 appears to be acceptable and has met the same criteria that a receiver gage would accept. Using the technique of two holes for inspection and measurement setups must give consideration to the holes that were made "zero" for establishing measurement planes to avoid rejection of acceptable parts and/or creation of unnecessary paperwork, rework, salvage, or scrap costs. The paper gaging technique uses a similar process found in most *best-fit centers software* of CMM devices. If your current process seems to reject parts that later assemble and work, you may wish to look at these current practices to see if they may be contributing to excessive scrap, rework, or salvage costs.

With this approach, all measurements and evaluation were performed regardless of feature (hole) size, or RFS. We did not take advantage of full available tolerances, nor consider bonus location tolerances available when the hole size departs from MMC size. These benefits are not available with the coordinate +/- dimensioning system.

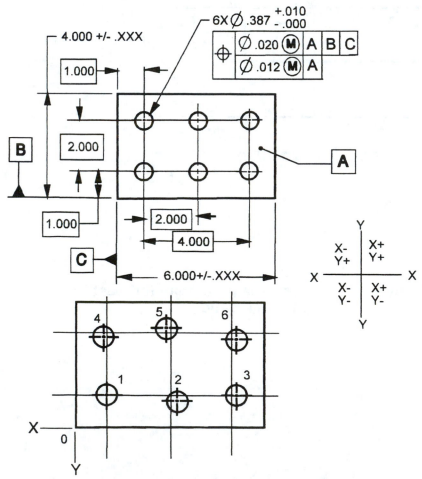

Figure 8-4   Paper gaging: positional tolerances.

PAPER GAGING AND POSITIONAL TOLERANCING

The part illustrated in Figure 8-4 is similar to Figure 8-3, but it has been dimensioned using position tolerance controls per ASME Y14.5M. The hole pattern is dimensioned from the part edges as before, but the edges are designated as datums, the tolerance zone is expressed as a diameter, and bonus tolerances are available due to the use of MMC. The pattern is located with a diameter tolerance of .020, whereas the hole-to hole tolerance is .012, both MMC. The datum reference frame specifies the pattern to be perpendicular to datum surface A. Using the same procedure as before, locate from the datum feature surfaces to establish the pattern location.

The next step, as shown in Figure 8-5, is to measure the location of each hole to determine its departure from true position and to establish the plot on the graph. Further, we need to measure the size of each hole to apply any available MMC size departure bonus tolerance.

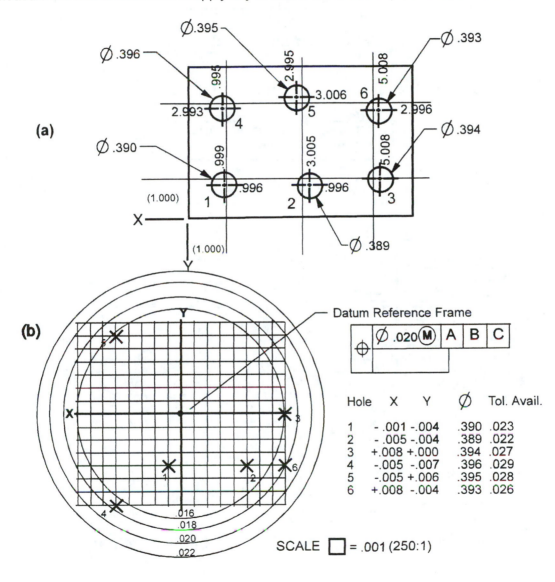

Hole	X	Y	$\varnothing$	Tol. Avail.
1	-.001	-.004	.390	.023
2	-.005	-.004	.389	.022
3	+.008	+.000	.394	.027
4	-.005	-.007	.396	.029
5	-.005	+.006	.395	.028
6	+.008	-.004	.393	.026

SCALE ▢ = .001 (250:1)

Figure 8-5 Paper gage plot: hole pattern location.

Next is the evaluation of the lower control frame, or hole-to-hole location tolerance, and the orientation of the pattern to datums specified. The tolerance zone is a diameter of .012, relative to datum A (perpendicularity). The holes are free to float within the upper control established .020 diameter tolerance zones. A bonus tolerance for each hole does not exist, because none were produced at MMC (.387 diameter). From the measurements, we determine that the holes are acceptable for size, and with an overlay to the same scale as the plot, we determine the holes are within location tolerance. See Figure 8-6. The process used for this figure is discussed next.

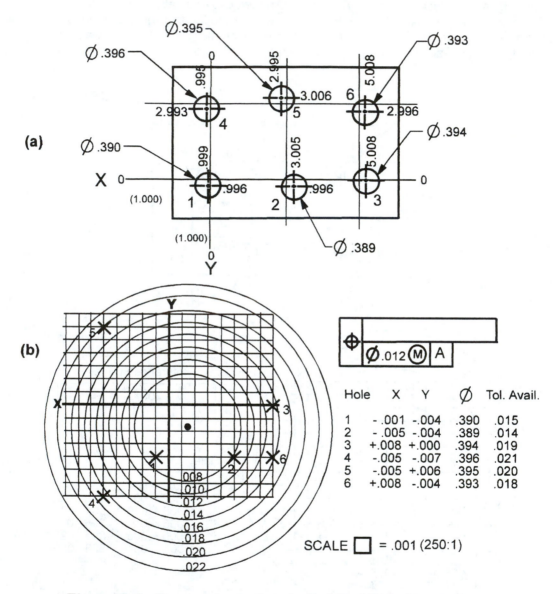

Hole	X	Y	Ø	Tol. Avail.
1	- .001	-.004	.390	.015
2	- .005	-.004	.389	.014
3	+.008	+.000	.394	.019
4	-.005	-.007	.396	.021
5	-.005	+.006	.395	.020
6	+.008	-.004	.393	.018

SCALE ☐ = .001 (250:1)

Figure 8-6   Paper gage plot: individual hole locations.

218

The scale has been established as 1/4 = .001 (250:1); observe the hole pattern plot. As hole 5 is displaced up and to the left, relative to the other holes of the pattern, hole 5 is considered as a worst case for the purpose of placement of the overlay in order to save time. The available hole tolerances have been determined and bonus tolerance applied. The overlay is placed with the .020 diameter tolerance circle tangent to the center of hole 5, ensuring that it is within tolerance. This will move the center of the overlay down and to the right .0015 (1-1/2 squares). This process accepts the same quality as a functional gage, allowing for size bonus tolerance and for the pattern location tolerance, as well as the hole-to-hole tolerance, to be independently verified. In addition, as the tolerance zones are diameters, there is an increase in tolerance zone area of 57% available to manufacturing.

EXERCISE ON PAPER GAGING: CIRCULAR PATTERN

We are now ready for an exercise in paper gaging. Figure 8-7 illustrates a circular pattern of six holes. The holes are located from datum framework B, A, C all RFS. The pattern is to be located from the datum frame within .010 diameter when the holes are at MMC size. In addition, the holes are to be relative to one another within .005 diameter MMC.

To locate the part for measurement, the part is to be flush to primary surface B, located on secondary datum diameter A, with rotation stopped by tertiary datum slot C. Complete this exercise using the overlay of Figure 8-10a.

1. Upper control callout: Using Figure 8-8, plot the hole locations, record hole size, and determine available tolerance and part acceptability.

2. Lower control callout: Using Figure 8-9, and the previous plot of locations, record size and determine tolerance available and part acceptability.

Review the example. Are the holes plotted correctly? Because the upper control specifies a position control relative to diameter A RFS, our overlay must be concentric with the axis of the datum frame and plot. The overlay cannot be adjusted or moved about. The part is acceptable for the upper control callout. Review the plot of holes for the lower control callout. Only datum B is specified, allowing the plot of holes to float within the upper tolerance limit but be perpendicular to surface B and relative to each other within .005 MMC. Because holes 1 and 6 are the farthest away of all the holes, we shall place the overlay on a line going through these holes. The best possible location for the overlay would be centered between holes 1 and 6. Are all holes within tolerance? No. Is the part scrap? No.

Notice that the hole size is .500 - .510 diameter. Holes 1 and 6 were made very close to the MMC size limit, allowing little bonus tolerance. If the holes were redrilled to a larger size, such as .506 diameter, what would the allowable tolerance be? Would the part be acceptable? Yes. Also note that this method has not allowed for the holes to be out of perpendicularity with surface B. We must consider this to accurately determine fit with the mating part and ensure full assembly. One option is to measure and plot both the entry and exit locations of each hole, if possible. Another is to consider equipment capability from quality audits and add that additional tolerance to each hole plot location. If the fixture and equipment are capable of holding perpendicularity of .002 per inch and if the part is .500 thick, we could add .001 to each hole plot, as an example of worst-case comparisons. CMMs, along with appropriate software, make this job much easier. Formulas for determining squareness of features may be found later in this chapter and in ASME Y14.5M, Appendix B.

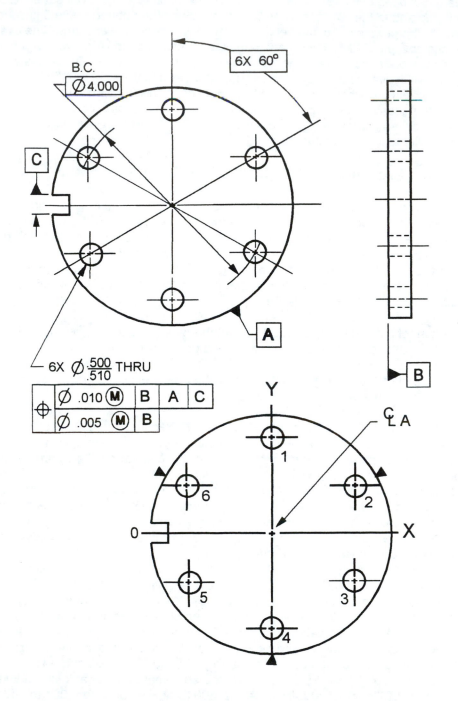

Figure 8-7 Paper gage: circular pattern.

220

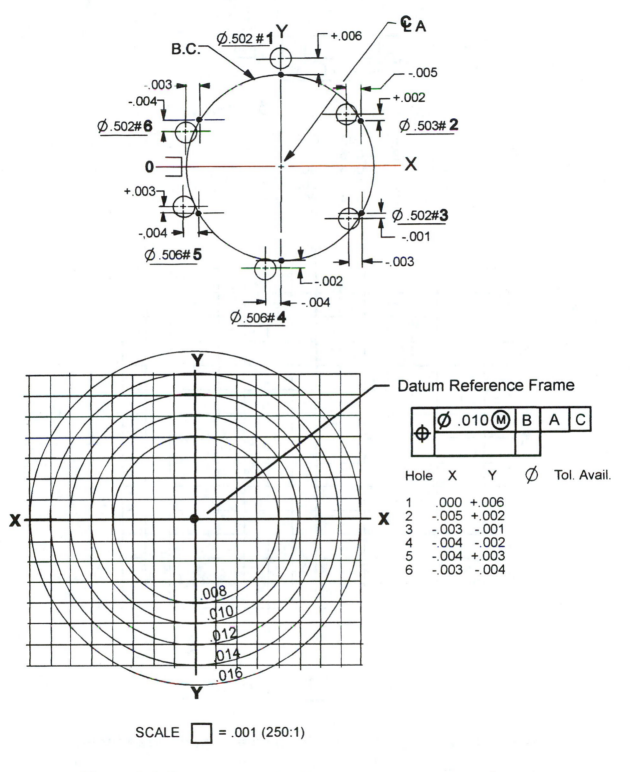

Figure 8-8 Paper gage: circular pattern, upper control.

Datum Reference Frame

⊕	Ø .010 Ⓜ	B	A	C

Hole	X	Y	Ø	Tol. Avail.
1	.000	+.006		
2	-.005	+.002		
3	-.003	-.001		
4	-.004	-.002		
5	-.004	+.003		
6	-.003	-.004		

SCALE ☐ = .001 (250:1)

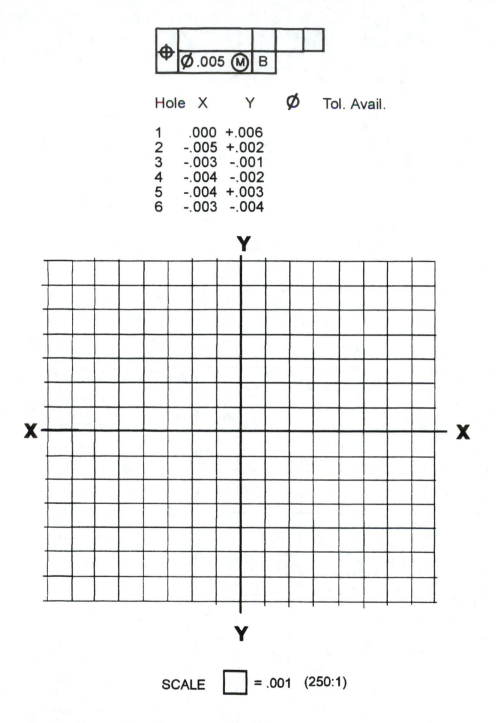

Hole	X	Y	Ø	Tol. Avail.
1	.000	+.006		
2	-.005	+.002		
3	-.003	-.001		
4	-.004	-.002		
5	-.004	+.003		
6	-.003	-.004		

SCALE ☐ = .001 (250:1)

Figure 8-9   Paper gage plot: individual hole location, lower control.

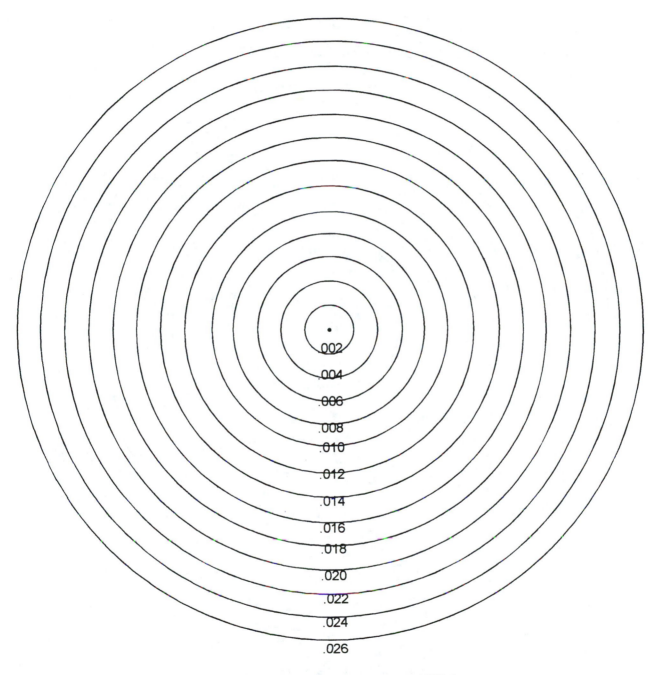

.002
.004
.006
.008
.010
.012
.014
.016
.018
.020
.022
.024
.026

(a) For inch measurements   250:1

Figure 8-10 Overlay.

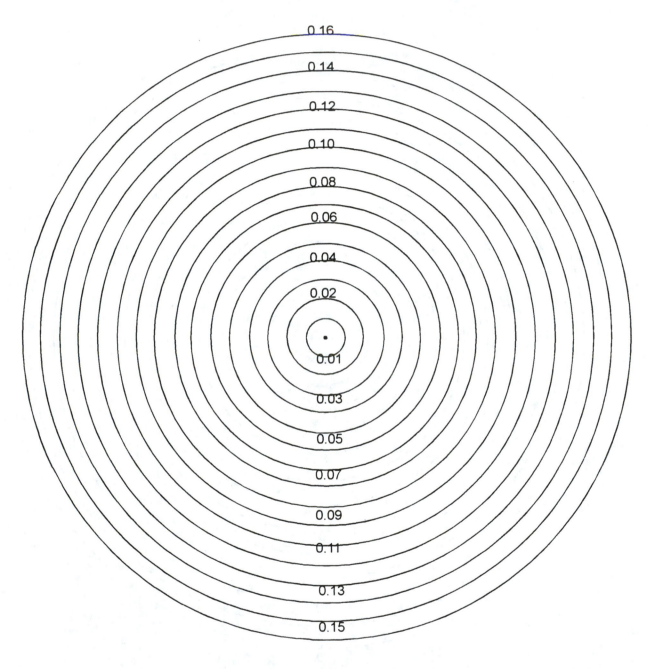

(b) For metric measurements   100:1

Figure 8-10 (continued)

224

FUNCTIONAL GAGING VERSUS OPEN SETUP INSPECTION

Not all products have sufficient volume to justify expense of dedicated functional gaging. When acceptance or rejection decisions are made, however, the same level of product quality must be maintained, regardless of the methods used. Figure 8-11 illustrates a part with a six-hole circular pattern relative to datum surface A primary and datum diameter B secondary. Notice that both the holes and the datum diameter have a tolerance, a position tolerance for the holes, and an orientation tolerance for the center pin diameter. Tolerances and the effects of MMC applied to features are not difficult to understand, but the application of datum tolerances and effects of bonus tolerances applied to datum features are sometimes misapplied.

Figure 8-12a shows the tolerance zones for features of a part made to LMC conditions, accepted by a functional gage. There is a tolerance of 0.4 diameter for datum axis B, when datum feature B is at LMC, and a tolerance of 1.0 diameter at each hole, when the holes are at LMC. The tolerance zones for the holes are free to move about the axis of the datum as influenced by the LMC conditions of both the datum feature and the feature holes.

Figure 8-12b illustrates an incorrect application of datum tolerance and bonus tolerances that would accept a part of greater tolerance at each hole axis, but would not consider shift of the pattern. A different level of product quality could be accepted by this process.

Figure 8-13a demonstrates the application of datums and tolerances for an open setup measurement that are comparable to the functional gage. Observe the results of both plots of Figure 8-13. Note in Figure 8-13a, the 1.0 diameter feature tolerance zone cannot accept both holes 1 and 6 simultaneously, but the 1.0 dia. tolerance zone is allowed to translate or shift about in the 0.4 diameter datum tolerance zone.

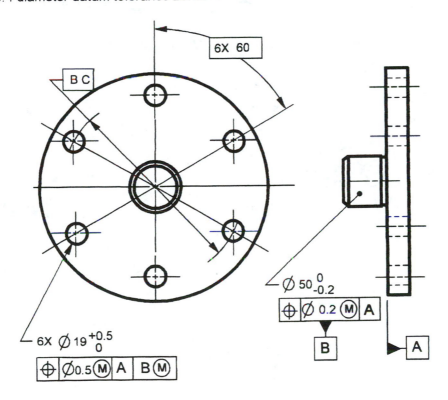

Figure 8-11 Position Tolerancing: gage versus open setup measurement.

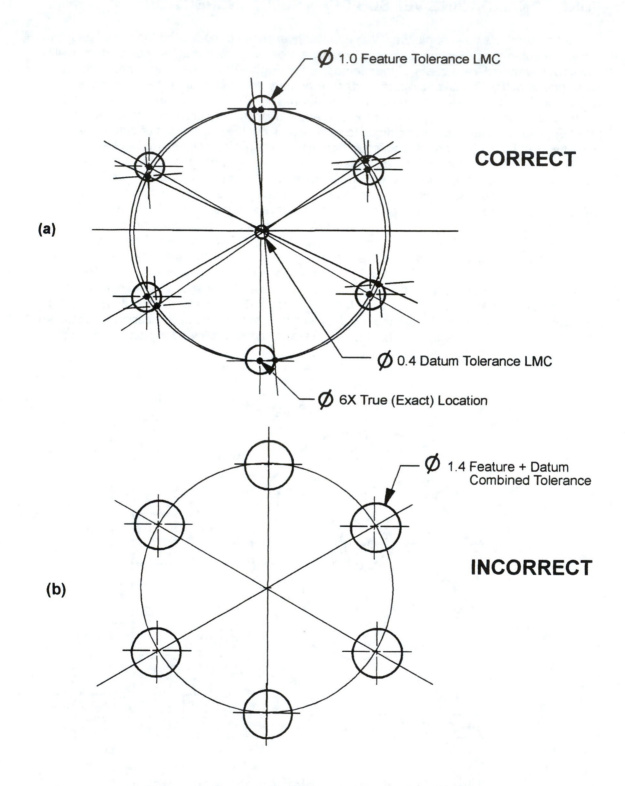

Figure 8-12 Feature and datum tolerance zones LMC.

The CMM/open setup paper gage technique
must accept the same level of workpiece
quality as the functional gage.

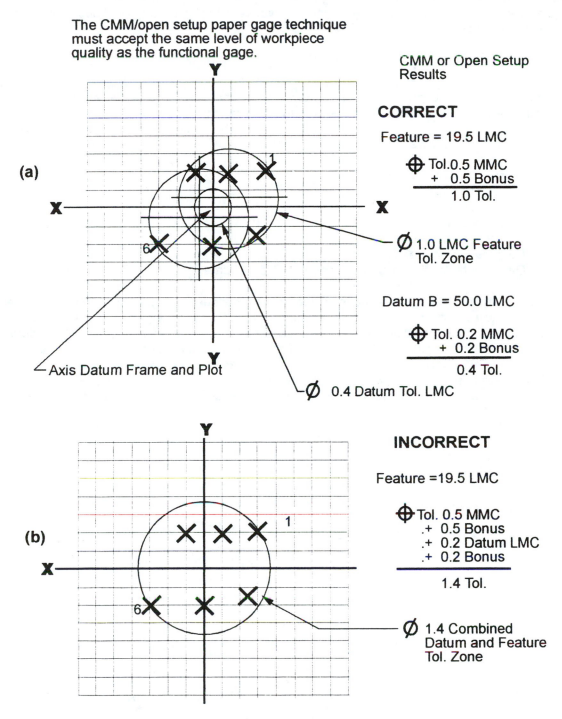

CMM or Open Setup
Results

CORRECT

Feature = 19.5 LMC

⊕ Tol. 0.5 MMC
    + 0.5 Bonus
    ───────────
       1.0 Tol.

∅ 1.0 LMC Feature
   Tol. Zone

Datum B = 50.0 LMC

⊕ Tol. 0.2 MMC
    + 0.2 Bonus
    ───────────
       0.4 Tol.

∅ 0.4 Datum Tol. LMC

─ Axis Datum Frame and Plot

INCORRECT

Feature = 19.5 LMC

⊕ Tol. 0.5 MMC
  .+ 0.5 Bonus
  .+ 0.2 Datum LMC
  .+ 0.2 Bonus
  ──────────────
       1.4 Tol.

∅ 1.4 Combined
   Datum and Feature
   Tol. Zone

Figure 8-13 Paper gage plot: datums and features LMC.

The lower plot has combined the feature tolerance and datum tolerance to arrive at a total tolerance level of 1.4 diameter, and has applied this number at each hole. Which method emulates the gage? Figure 8-13a rejects the part, whereas Figure 8-13b would accept it! This part is out of spec!

PAPER GAGING SUMMARY

Paper gaging, although not the answer for every situation, offers certain advantages that should be carefully evaluated for each design and application:

Immediate analysis
Low cost
Independent of product quantities
No required gage design
No lead time
Provides some process data
No gage wear allowances
No gage storage requirements
Offers more precise inspection reports and setup data

In summary, paper gaging is not dedicated or fixed, because it may be applied to prototype parts or very low volume parts, including audits. Performed properly, it gives information from a controlled setup, thus providing information on process data, fixture accuracy, and drill spindle and/or punch wear. It affords the benefits of immediate analysis and feedback for process control and supplier audits. Costs for gage design development and storage may be avoided.

Paper gaging is not the answer for every situation, but given proper thought, its use can complement a quality system, providing an extra tool for engineering, manufacturing, and quality.

9 OTHER CONTROLS, CONVENTIONS AND SYMBOLS

DIMENSION ORIGIN SYMBOL

The *dimension origin* symbol was added to the 1982 Y14.5 standard to indicate the origin of a dimension, when it was necessary to clarify on drawings where a tolerance zone was to exist. Drawings made from former practices did not always have datum references.

Figure 9-1 illustrates a simple part, and because it is difficult to determine if the tolerance exists at the foot plane or at the hole axis, this symbol helps clarify the design intent. When added to older drawings, this symbol must carry a note of clarification as to where the symbol came from. Because drawings and standards exist for decades, and because the 1982 standard did not exist when some active drawings were made, the clarification is mandatory. Such a note might read, "PER ASME Y14.5M-1994."

The intent of this symbol is not to replace the need for datum references, but it does offer a quick fix on older drawings where development of datum frames and extensive change is prohibited.

HOLE DEPTH DIMENSIONING

One possible application of the dimension origin symbol is for dimensioning hole depths in curved or warped surfaces. These features are often difficult to evaluate with conventional gaging methods. When holes are produced in curved surfaces, the dimensioning and tolerancing method shown in Figure 9-2 may offer some benefits. This technique allows a gage pin of given length to be inserted in the hole, a measurement to be taken from the opposite side shaft OD, and the pin length to be subtracted from the measurement to determine the produced depth. This principle is similar to that used on shaft keyslots and keyways.

PROJECTED TOLERANCE ZONE

The *projected tolerance* concept "projects" the normal position or perpendicularity tolerance zone above or below the surface indicated on the drawing. The application of a projected tolerance zone is recommended where the variation in perpendicularity of the feature (threads or force fit diameter) could allow fastener interference in mating parts. An interference can occur where a press fit pin or threaded fastener is inclined within the positional tolerance limit, as shown in Figure 9-3. Unlike the *floating fastener* applications involving all clearance holes, the attitude of the *fixed fastener* is governed by the accuracy of the produced hole into which it assembles. Methods of projected tolerance zone symbol application are shown in Figure 9-3. When the projected tolerance symbol is not used, it will be necessary to select a clearance hole size that will ensure a clearance fit. The formula from Y14.5 to allow for this condition is

$$H = F + T1 + T2 \left(1 + \frac{2P}{D}\right).$$

where T1 = position tolerance diameter of clearance hole.
   T2 = position tolerance diameter of threaded or press fit hole.
   D = depth of threaded or press fastener.
   P = maximum fastener projected height.
   H = clearance hole size.
   P = fastener diameter.

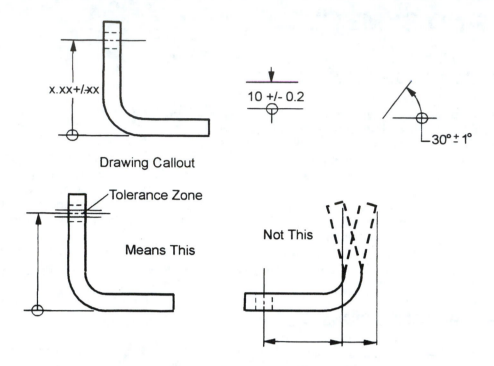

Drawing Callout

Tolerance Zone

Means This

Not This

Figure 9-1 Dimension origin symbol.

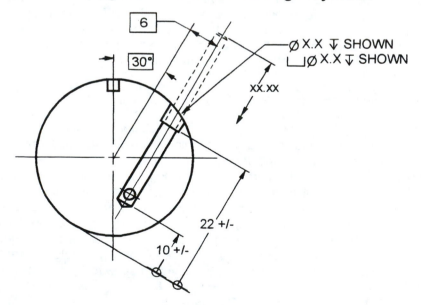

Figure 9-2 Hole depth dimensioning: curved surfaces.

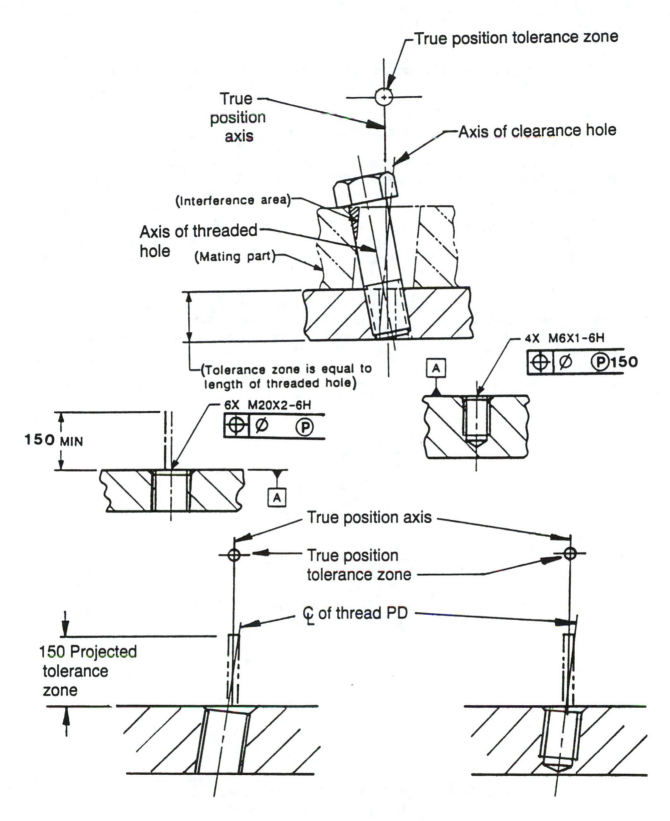

Figure 9-3 Projected tolerance zone.

TAPER AND SLOPE

A *slope* is the inclination of a surface expressed as a ratio of the difference in heights at each end (above and at right angles to a base line) to the distance between those heights.

A *taper* is defined as the ratio of the difference in the diameters of two sections (perpendicular to the axis) of a cone to the distance between those sections.

A *conical taper* includes tapers defined in ANSI B5.10 standard machine tapers used throughout the tooling industry that are classified as *American Standard Self-Holding and Steep Taper series*.

The *self-holding series* has 22 sizes: three smaller sizes with a taper of 1/2 inch per foot, eight sizes with a taper of 5/8 inch per foot, and eleven sizes with a taper of 3/4 inch per foot. The *steep taper series* has twelve sizes and are defined as having a sufficiently large angle as to ensure an easy self-releasing feature.

American Standard Machine Tapers are generally dimensioned using the taper name and number, with reference to the taper per foot. The diameter at the gage line, as well as length is sometimes given, with the small taper end as reference. See Figure 9-4a and b.

Taper and Slope may also be specified by the method illustrated in Figure 9-5. For further information on machine tapers, refer to ANSI B5.10.

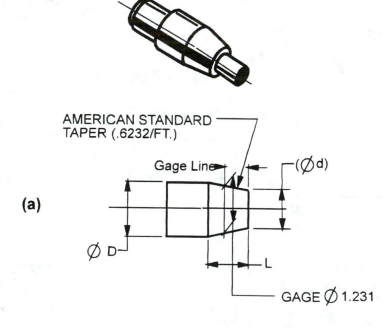

(a)

AMERICAN STANDARD TAPER (.6232/FT.)

Gage Line

(Ø d)

Ø D

L

GAGE Ø 1.231

Figure 9-4   (a) American Standard Machine Taper; (b) taper features and application.

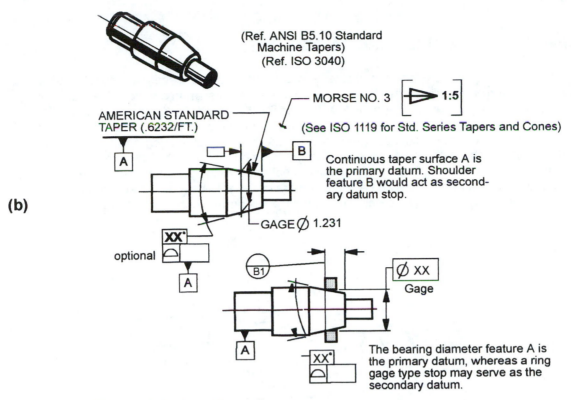

(Ref. ANSI B5.10 Standard
Machine Tapers)
(Ref. ISO 3040)

MORSE NO. 3 ▷ 1:5

AMERICAN STANDARD
TAPER (.6232/FT.)

A

B

(See ISO 1119 for Std. Series Tapers and Cones)

Continuous taper surface A is
the primary datum. Shoulder
feature B would act as secondary datum stop.

(b)

GAGE ⌀ 1.231

optional

XX°

A

B1

⌀ XX
Gage

A

XX°

The bearing diameter feature A is
the primary datum, whereas a ring
gage type stop may serve as the
secondary datum.

Figure 9-4 (continued)

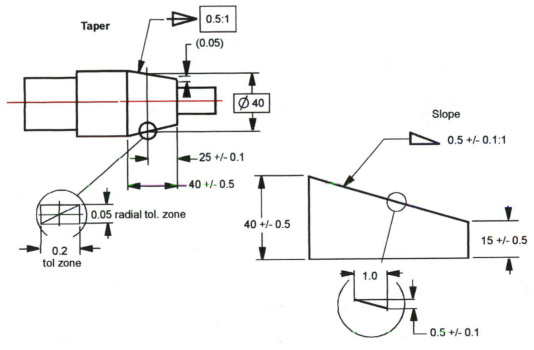

Taper

0.5:1

(0.05)

⌀ 40

25 +/- 0.1

40 +/- 0.5

0.05 radial tol. zone

0.2
tol zone

Slope

0.5 +/- 0.1:1

40 +/- 0.5

15 +/- 0.5

1.0

0.5 +/- 0.1

Figure 9-5 Taper and slope.

SPHERICAL FEATURES AND DATUMS

Spherical features are features of size; they are continuous surfaces with possible size, form, and location error. *Profile* is often used to refine the errors of size. In addition, the feature may require a location tolerance control MMC, LMC, or RFS. Figure 9-6 illustrates a spherical feature of size with a tolerance of 40.0 - 39.5 diameter size, and with a profile tolerance refinement tolerance of 0.2 that must exist between points X and Y. The profile tolerance exists for the full feature surface, therefore the *surface profile* symbol, and must lie within the size tolerance. The spherical surface has a location tolerance of 0.4 spherical diameter relative to primary datum B RFS and secondary datum surface A. Note the use of the spherical diameter symbol in the feature control frame. To ensure design intent, note 1 is added to the drawing controls.

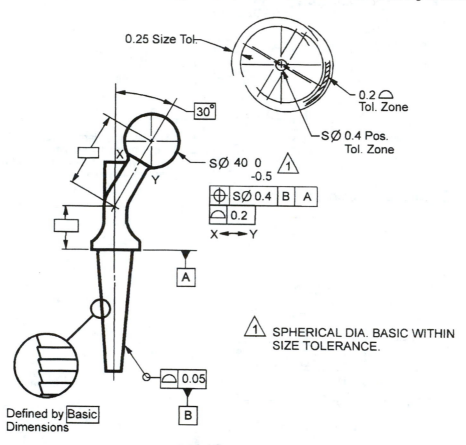

Figure 9-6   Spherical features: size, form, and location control.

If spherical features are chosen as datums, it may become difficult to stabilize the part for measurement. Figure 9-7 illustrates a shaft with a spherical ball end used as the primary datum feature. With only datum A invoked, the part is free to rotate in space as shown in view (a). If secondary line contacts were to be made at selected points of the shaft diameter, as shown in view (b), one more degree of freedom is removed; however, the shaft may still rotate from side to side. By adding a second line contact to the shaft surface as shown in view (c), the side-to-side movement is arrested, but axial rotation is still possible because the primary datum A is only a point. In these cases, if it is determined that the primary datum must be a spherical feature (thus a point), a tertiary datum slot, or a flat may be introduced to stop all rotations.

234

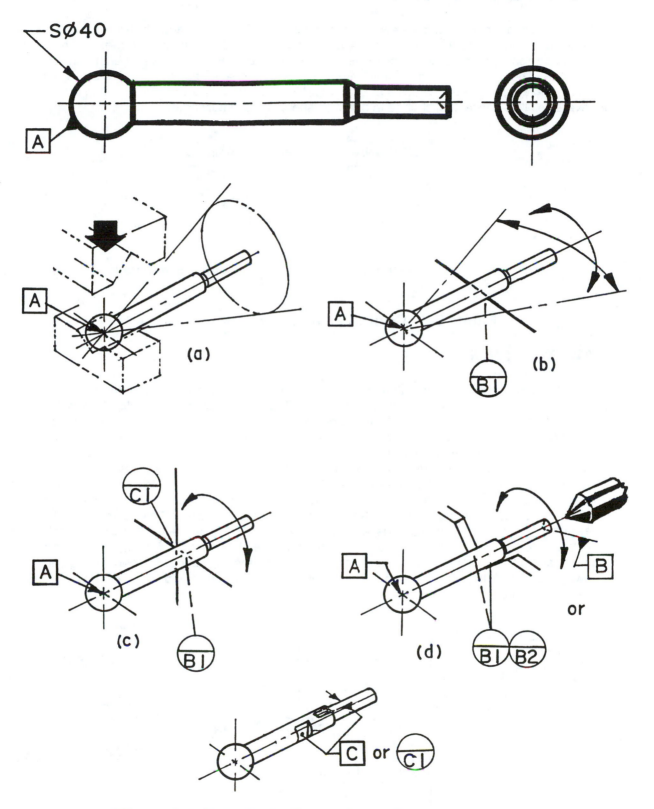

Figure 9-7 Spherical primary datum features.

CONTROLLED RADIUS SYMBOL

Controlled radius (CR) is a new concept in the 1994 Y14.5 standard that allows a variable-size radius, but without flats or reversals, as shown in Figure 9-8. A controlled radius allows a tolerance zone defined by two arcs (minimum and maximum radii) that are tangent to adjacent surfaces. The feature contour must lie within the crescent-shaped zone, a true fair curve, without flats or reversals. In the previous standard, ANSI Y14.5M-1982, the symbol R implied no flats or reversals. This conceptual change in direction should be carefully noted.

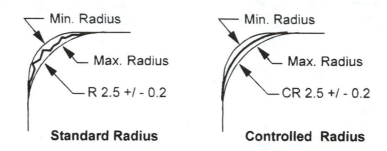

Figure 9-8   Controlled and standard radius.

AVERAGE DIAMETER

Rule 1 states that *no element of a feature shall extend beyond the perfect form size limits at MMC*. This rule has a disclaimer relative to *nonrigid parts*, such as rubber seals, O-rings, fibrous material products, thin section parts like large-diameter ring gears, or other metallic products that may be subject to process distortions such as heat treatment. For many products, it may not be possible to control them in the free state referenced in Rule 1.

The use of the *average diameter* concept may be appropriate for these products. Average diameter is used when it is necessary to control a feature to a more liberally defined shape, knowing that the parent or mating parts of the next level assembly will contain the shape to a more refined form. Average diameter is the average of multiple diametral measurements of a feature in its *free state*. When necessary, the average diameter may be found by peripheral measurement. The free state measurements(s) may exceed the size tolerance, but the *average diameter* must fall within the tolerance specified. Further, the difference between the high and low measured limits cannot exceed the *circularity control* value specified. The *free state symbol* is applied in the circularity control frame as shown in Figure 9-9.

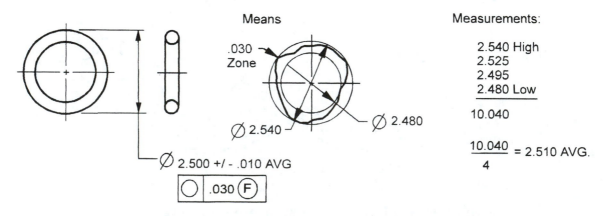

Figure 9-9   Average diameter and free state controls.

Free state variation describes part distortions when removed from restraints imposed by the manufacturing process. These distortions may be due to weight and flexibility of the part or to internal stresses imposed by fabrication. Thin walls in proportion to size, such as jet engine housings, are another example. The term used in the previous paragraphs is *nonrigid parts*. Figure 9-10 illustrates the use of the average diameter method of dimensioning along with the free state symbol and a note to explain restraining forces needed for the measurement process to emulate the assembly process. Without note 1, all specifications would apply in the free, unrestrained state.

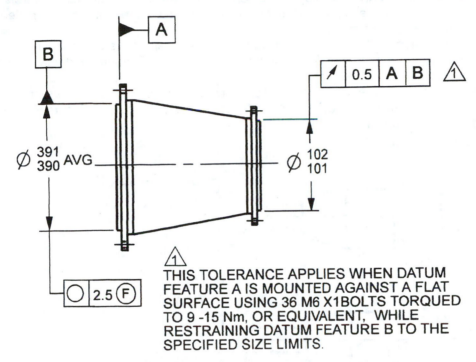

⚠1 THIS TOLERANCE APPLIES WHEN DATUM FEATURE A IS MOUNTED AGAINST A FLAT SURFACE USING 36 M6 X1BOLTS TORQUED TO 9 -15 Nm, OR EQUIVALENT, WHILE RESTRAINING DATUM FEATURE B TO THE SPECIFIED SIZE LIMITS.

Figure 9-10 Free state variation: restraint of nonrigid parts.

STATISTICAL TOLERANCE CONTROLS

Statistical tolerancing generally has not been applied to product drawings, because it is considered to be in the realm of *process or manufacturing engineering* information. The 1994 ASME Y14.5 standard has recognized the need for a universal symbol and drawing technique as a recommendation when statistical tolerancing specifications are included on a drawing. As a personal preference, I use *statistical tolerance controls* only:

1. After exhausting direct tolerancing, 100% fit methods
2. As a joint design/manufacturing/quality effort
3. When multiplant/supplier issues are minimal
4. When statistical process control documentation is in place

Statistical tolerancing involves assigning tolerances to related components of an assembly on the basis that the assembly tolerance is equated to the square root of the sum of the squares of the individual component tolerances. When 100% fit tolerancing becomes overly restrictive, or nearly improbable for manufacturing, statistical tolerancing may be used. Increasing tolerances by this method may reduce manufacturing costs, but should only be employed where appropriate process controls are in place. See Figure 9-11a. Occasionally, when the dimension and tolerance has a possibility of being produced without statistical process control (SPC), it may be desired to place both the arithmetical design limit and the statistical tolerance limit on the drawing. See Figure 9-11b.

Statistical tolerancing principles may also be applied to geometric tolerancing controls, as shown in Figure 9-12. In this example, statistical tolerancing techniques have been applied to both the size and the position control.

Generally, the use of statistical tolerancing should be a collaborative process, with tolerances arrived at through the involvement and input from manufacturing and quality. Process control systems may vary between companies or industries, but when shown as part of the drawing specification, they are part of the contract between supplier and customer and should be documented via a process control document or standard. The note shown in Figure 9-11 is also recommended for each drawing.

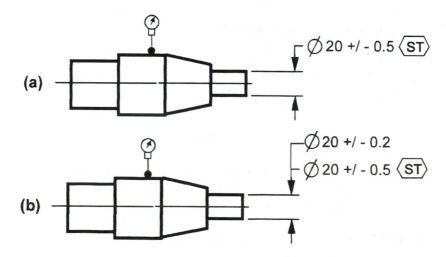

NOTE:
FEATURES IDENTIFIED AS STATISTICALLY TOLERANCED ⟨ST⟩
SHALL BE PRODUCED WITH STATISTICAL PROCESS CONTROLS
OR TO THE MORE RESTRICTIVE ARITHMETIC LIMITS.

Figure 9-11   Statistical tolerance symbol.

VIRTUAL AND RESULTANT CONDITION

Virtual and resultant conditions discussed earlier are mentioned again in the definition section as well as the glossary. A diagram helps explain and clarify the concepts.

MMC and LMC are opposites as well as *inner and outer boundaries*. Virtual conditions may occur at MMC or LMC. MMC will give the closest fit tolerances with mating parts, whereas LMC will utilize the maximum material. Resultant conditions contain the additive effects of all size, geometric, and bonus tolerances. Figure 9-13 is a diagram for either holes or shafts, at MMC or LMC, with applicable bonus tolerance applied. Review this diagram, noting the effects of MMC/LMC and bonus tolerance, as applicable.

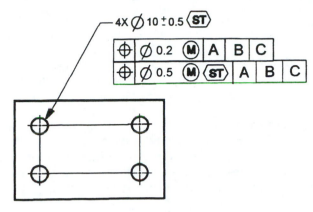

Figure 9-12 Statistical tolerancing: geometric controls.

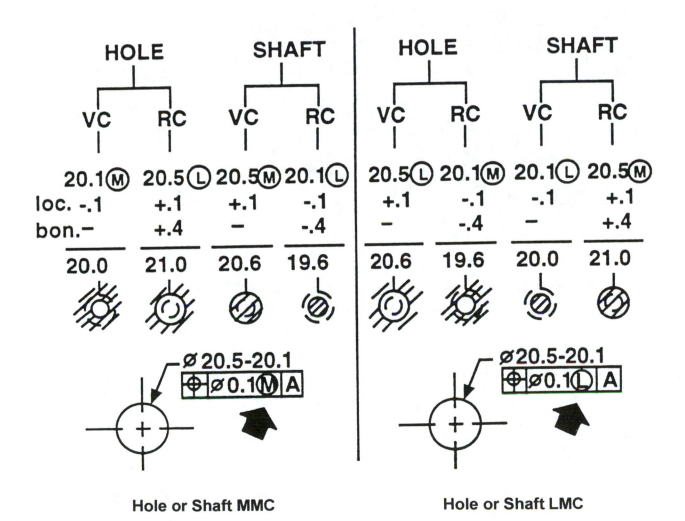

Hole or Shaft MMC   Hole or Shaft LMC

Figure 9-13 Virtual and resultant condition diagram.

SURFACE TEXTURE

Surface texture is covered by ANSI B46.1 standard, and although this subject is not part of the Y14.5 standard, the specified tolerances on engineering drawings have a direct relationship to feature surface texture, sometimes referred to as primary and secondary texture. This relationship is due to the interrelationship of equipment/process capability and specified tolerances. Figure 9-14 illustrates this relationship and how tolerances, processes, and resulting surface textures are interlinked. Generally, it would be impractical to specify a size tolerance of .010 inch, with primary surface finish mark of 8 microinches or a form tolerance of .0004 with a surface finish of 250 microinches. There may be exceptions, but design tolerances specified will generally have a direct impact on equipment capability, and visa versa.

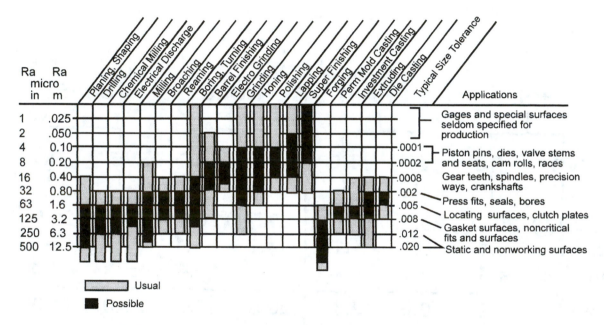

**Figure 9-14   Surface texture versus tolerances
and production processes.**

Information relative to surface texture may be found in ANSI B46.1, with drawing practice and symbology covered in ANSI/ASME Y14.36. A comparison of microinch, micrometer, and standard ISO roughness grades is shown in Figure 9-15.

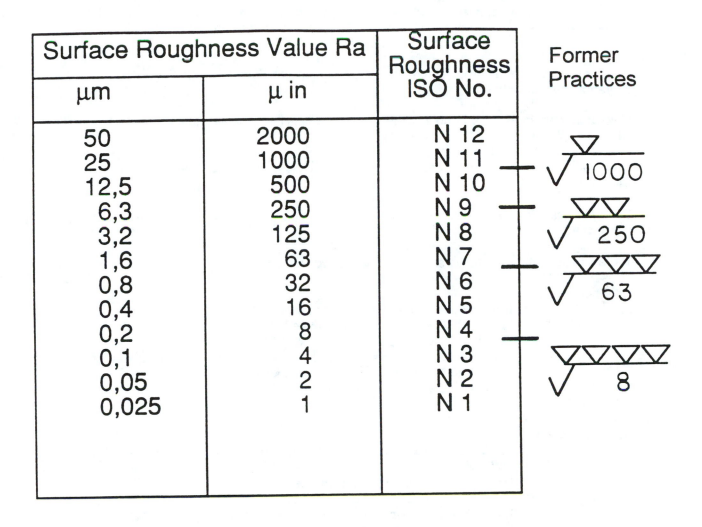

Surface Roughness Value Ra		Surface Roughness ISO No.	Former Practices
μm	μ in		
50	2000	N 12	
25	1000	N 11	
12,5	500	N 10	
6,3	250	N 9	
3,2	125	N 8	
1,6	63	N 7	
0,8	32	N 6	
0,4	16	N 5	
0,2	8	N 4	
0,1	4	N 3	
0,05	2	N 2	
0,025	1	N 1	

∨ Surface may be produced by any method

▽ Material removal by machining is required; material must be provided for that purpose

▽ Material removal allowance in inches (or millimeters)

∀ Material removal prohibited

MAX. WAVINESS HEIGHT
MAX WAVINESS WIDTH
MAX. Ra —▸63
.002-2
MIN. Ra —▸32
.030◂ CUTOFF
LAY
⊥.015◂ MAX. ROUGHNESS SPACING

Figure 9-15 Surface texture standard values.

INNER AND OUTER BOUNDARY

The terms *Inner and Outer Boundary* are broad terms to describe worst-case fits and stackups and are applied to features RFS, MMC, or LMC. Virtual and resultant condition terms, however, are only applied to features at MMC or LMC conditions. The following definitions will help in determination of the *boundary:*

Inner boundary for an internal feature (hole) is the smallest hole minus the geometric tolerance (if specified).

Outer boundary for the above hole is the largest hole plus the geometric tolerance (if specified).

Inner boundary for an external feature (shaft) is the smallest shaft minus the geometric tolerance (if specified).

Outer boundary for the above shaft is the largest shaft plus the geometric tolerance (if specified).

The above definitions state *worst-case conditions,* which mean:

Inner boundary:

Holes	MMC = MMC size - position tolerance MMC
	LMC = MMC size - position tolerance LMC - bonus position tolerance
	RFS = MMC size - position tolerance RFS
Shafts	MMC = LMC size - position tolerance MMC - bonus position tolerance
	LMC = LMC size - position tolerance LMC
	RFS = LMC size - position tolerance RFS

Outer boundary:

Holes	MMC = LMC size + position tolerance MMC + bonus position tolerance
	LMC = LMC size + position tolerance LMC
	RFS = LMC size + position tolerance RFS
Shafts	MMC = MMC size + position tolerance MMC
	LMC = MMC size + position tolerance MMC + bonus position tolerance
	RFS = MMC size + position tolerance RFS

Note: Orientation could be used in place of position in these formulas.

APPENDIX

FEATURE CONTROL DIAGRAM

Figure A1 is intended as an aid for determining which geometric controls ARE most appropriate. The diagram is structured with yes or no questions that search for appropriate answers. Starting at the top, follow the chart until your question or requirement is resolved. For complex designs, it will be necessary to go through the chart for each issue in question. The diagram may not satisfy every issue or design circumstance, but with reasonable care, most of your design and/or drawing issues should be covered. ASME Y14.5M-1994 has a similar, more complex set of diagram figures, should you wish to go further.

QUALITY CONTROL PLANS

The quality function is primarily geared toward the activities of a product's acceptance or rejection and gathering/analyzing variables data for process control or other reasons. As products are developed, so are the quality plans for those products. Two key elements that greatly impact these plans are the plant environment (or personality) and the available measurement tools and equipment. A typical menu of issues is shown below.

Plant Environment
Projected Volume and Measurement Frequency
Accuracy Required
Product Risk Exposure: Employees, Customer, Public
Equipment and Process:Maintenance Schedule
Warranty Records and Past Product History
Personnel Systems and Training
People Needs
Other Issues

Measurement/Gaging Options
Audits: No Gaging
General Purpose
Optical Comparators, Polar/Linear Charts
Transparent Overlays
Air/Electronic
Laser Assist
Dedicated Gages
Surface Plate Inspection
CMM Inspection
Paper Gaging and Mathematical Analysis

Though not all inclusive, quality plans will be influenced by this menu, and these issues must be addressed, in a joint effort by engineering, manufacturing, and quality. See Figure A2. The order of questions may differ in priority relative to your company or organization, but I believe that the same (or similar) issues will arise and must be resolved. An ideal situation is one where design change is nil, and the process is backed by skilled personnel in all positions, with a full range of measurement/gaging options from which to choose.

As you work through the diagram, are there issues you would like to include? How would you reorder the questions?

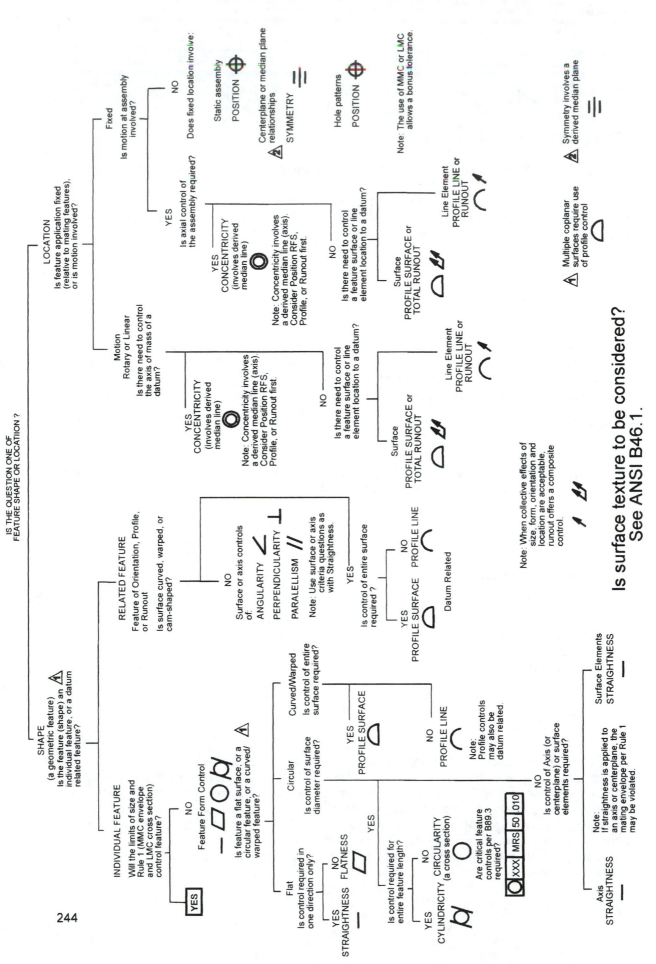

244

Figure A1 GDT tolerance control diagram.

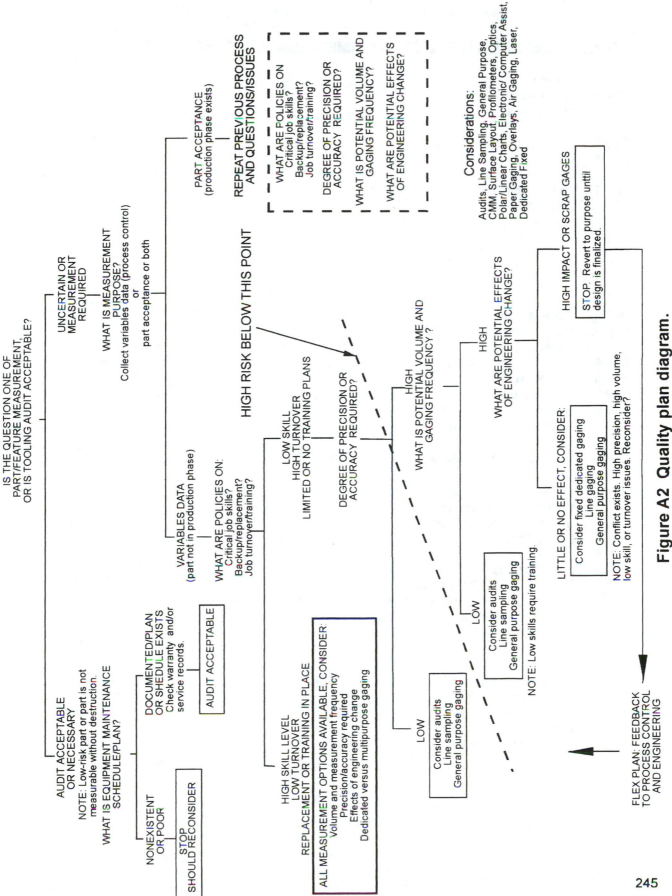

Figure A2 Quality plan diagram.

245

ASME STANDARDS RELATED TO DIMENSIONAL METROLOGY

B89.1   Length
B89.1.2   Gage Blocks and
          Measurements
B89.1.5   Circular Gages - External
B89.1.6   Circular Gages - Internal
B89.1.8   Lasers
B89.1.10 Dial Indicators
B89.1.13 Internal Micrometers and Calipers
B89.1.17 Thread Wires
B89.2   Angles
B89.3   Geometry
B89.3.1   Roundness (Circularity)
B89.3.2   Measurement Methods
B89.3.3   Flatness/Straightness
B89.3.4   Axes of Rotation
B89.3.5   Surface Plates
B89.3.6   Gaging and Fixturing GDT
          (replacing B4.4)
B89.4   CMM Technology
B89.5   Metrology Terms
B89.6   Environment
B89.6.1   Vibration
B89.6.2   Temperature and Humidity

OTHER RELATED ASME/ANSI STANDARDS

B1.1   Unified Screw Threads
B1.2   Gaging for UN Threads
B1.3   Screw Thread Acceptance
B1.13   Metric Threads
B1.15   UNJ Threads
B1.20   Pipe Threads
B4.1   Preferred Inch Limits and Fits
B4.2   Preferred Metric Limits and Fits
B4.4   Inspection of Workpiece (B89.3.6)
B5   Machine Tools Series
B18   Fasteners Series
B46   Surface Qualities
B46.1   Profile Methods
B46.4   Correlation
B47   Gage Blanks
B94   Cutting Tools, Holders, Drivers,
          and Bushings Series

246

AMERICAN NATIONAL DRAWING STANDARDS

ASME Y14.1M	DRAWING SHEET SIZE AND FORMAT
ASME Y14.2	LINE CONVENTIONS AND LETTERING
ASME Y14.3M	MULTISECTIONAL VIEW DRAWINGS
ASME Y14.4M	PICTORIAL DRAWINGS
ASME Y14.5M	DIMENSIONING AND TOLERANCING
ASME Y14.5.1M	MATHEMATICAL DEFINITIONS FOR Y14.5M
ASME Y14.5.2	CERTIFICATION OF GDT PROFESSIONALS (new)
ANSI Y14.6	SCREW THREAD REPRESENTATION
ANSI Y14.7	GEARS AND SPLINES
ASME Y14.8	CASTINGS AND FORGINGS
ANSI Y14.11	MOLDED PARTS DRAWINGS
ANSI Y14.13	MECHANICAL SPRING REPRESENTATION
ANSI Y14.17	FLUID POWER DIAGRAMS
ASME Y14.18M	OPTICAL PARTS DRAWINGS
ASME Y14.24M	TYPES AND APPLICATION FOR ENGINEERING DRAWINGS
ANSI Y14.31	UNDIMENSIONED DRAWINGS
ASME Y14.32.1	CHASSIS FRAMES
ASME Y14.32.2	BODY DRAFTING (new)
ASME Y14.35M	REVISION of ENGINEERING DRAWINGS (new)
ANSI Y14.36	SURFACE TEXTURE SYMBOLS
ASME Y14.5.37	COMPOSITE PARTS DRAWINGS
ASME Y14.38	ABREVIATIONS (new)
ASME Y14.39	LIMITS AND FITS (new)
AMSE Y14.40	GRAPHIC SYMBOLS (new)
ASME Y14.100	MIL-STD 100 (proposed)
ANSI/IEEE 268	METRIC PRACTICE

METRIC/IMPERIAL UNIT CONVERSION

Quantity	Multiply	By or Use Formula	To Obtain Equivalent Number of:	
acceleration	ft/sec^2	0.304 8	$\mathbf{m/s^2}$	(metre/second squared)
	in/sec^2	0.025 4	$\mathbf{m/s^2}$	
area	ft^2	0.092 9	$\mathbf{m^2}$	(square metres)
	in^2	6.452×10^{-4}	$\mathbf{m^2}$	
bending moment or torque	kg$_f$–m	9.807	$\mathbf{N \bullet m}$	(newton • metre)
	oz$_f$–in	7.062×10^{-3}	$\mathbf{N \bullet m}$	
	lb$_f$–in	0.113	$\mathbf{N \bullet m}$	
	lb$_f$–ft	1.356	$\mathbf{N \bullet m}$	
density	lb$_m$/ft^3	16.03	$\mathbf{kg/m^3}$	(kilogram/cubic metre)
	lb$_m$/in^3	27.70×10^3	$\mathbf{kg/m^3}$	
	lb$_m$/gal (U.S.)	0.119 9	$\mathbf{kg/}l$	
	lb$_m$/gal (Imp.)	0.099 9	$\mathbf{kg/}l$	
energy, work and heat	btu	1.055	$\mathbf{J}$	(joule)
	lb$_f$–ft	1.356	$\mathbf{J}$	
flow meter	cts/gal (U.S.)	0.264	$\mathbf{cts/}l$	(counts/litre)
	cts/gal (Imp.)	0.220	$\mathbf{cts/}l$	
	cts/ft^3	3.531×10^{-2}	$\mathbf{cts/}l$	
flow rate	lb$_m$/hr	1.26×10^{-4}	kg/s	(kilogram/second)
	lb$_m$/hr	0.454	$\mathbf{kg/hr}$	(kilogram/hour)
	*gm/hr	2.778×10^{-7}	kg/s	
	*gm/min	1.667×10^{-5}	kg/s	
			*gm/hr and gm/min are preferred units requiring no conversion.	
force	kg$_f$	9.807	N	(newton)
	kilopond$_f$	9.807	N	
	lb$_f$	4.448	N	
	poundal$_f$	0.138 3	N	
			up to 1000 lb$_f$ use N (newton) 4.448	
			from 1000–100,000 lb$_f$ use kN (kilonewton) 4.448×10^{-3}	
			above 100,000 lb$_f$ use MN (meganewton) 4.448×10^{-6}	
fuel performance	miles/gal (U.S.)	0.425 1	$\mathbf{km/}l$	(kilometre/litre)
	miles/gal (Imp.)	0.353 9	$\mathbf{km/}l$	
	miles/qt (U.S.)	1.701	$\mathbf{km/}l$	
	miles/qt (Imp.)	1.415 6	$\mathbf{km/}l$	
	gal/mile (U.S.)	2.352	$l\mathbf{/km}$	(litre/kilometre)
	gal/mile (Imp.)	2.825	$l\mathbf{/km}$	
	lb$_m$/bhp•hr	0.608×10^{-3}	kg/W•hr	(kilogram/watt•hour)
	lb$_m$/bhp•hr	0.608	$\mathbf{kg/kW \bullet hr}$	(kilogram/kilowatt•hour)
length	ft	0.304 8	$\mathbf{m}$	(metre)
	in	25.4	$\mathbf{mm}$	(millimetre)—all engineering drawings will be dimensioned in mm
	mile	1.609	$\mathbf{km}$	(kilometre)
mass	oz (avoir)	0.028 4	$\mathbf{kg}$	(kilogram)
	lb (avoir)	0.454	$\mathbf{kg}$	
	slug	14.6	$\mathbf{kg}$	
moment of inertia (2nd moment of area)	lb$_m$–ft^2	4.217×10^{-2}	$\mathbf{kg \bullet m^2}$	(kilogram•metre squared)
	lb$_m$–in^2	2.929×10^{-4}	$\mathbf{kg \bullet m^2}$	
power and heat rejection	hp	746	$\mathbf{W}$	(watt)
	btu/min	17.58	$\mathbf{W}$	
			below 1 hp use W (watt) 746	
			above 1 hp use KW (kilowatt) 0.746	

248

METRIC/IMPERIAL UNIT CONVERSION

Quantity	Multiply	By or Use Formula	To Obtain Equivalent Number of:		
pressure	in Hg	25.4	**mm Hg**	(millimetre of mercury)	
	in Hg @ 32°F	3.386×10^3	Pa	(pascal)	
	in Hg @ 60°F	3.377×10^3	Pa		
	in H$_2$O	25.4	**mm H$_2$O**	(millimetre of water)	
	in H$_2$O @ 39.2°F	249	Pa		
	in H$_2$O @ 60°F	248.84	Pa		
stress	kg$_f$/mm^2	9.8×10^6	**Pa**		
	lb$_f$/ft^2	47.88	**Pa**		
	lb$_f$/in^2(psi)	6.894×10^3	***Pa**		
	poundal/ft^2	1.488	**Pa**		

			*up to 1 psi, use Pa (pascal) 6.894×10^3
			  from 1—1000 psi, use kPa (kilopescal) 6.894
			  above 1000 psi, use MPa (megapascal) 6.894×10^{-3}		

temperature	**°Celsius**	to$_C$ + 273.15	K	(kelvin)	
	°Fahrenheit	(to$_F$ + 459.67)/1.8	K		
	°Rankine	t$_R$/1.8	K		
	°Fahrenheit	(to$_F$ −32)/1.8	**°C**	**(Celsius) (Centigrade)**	
	°Kelvin	t$_K$ −273.15	°C		
temperature interval	**°Celsius**	1.0	K		
	°Fahrenheit	5.556×10^{-1}	K or °C		
velocity	ft/min	5.08×10^{-3}	**m/s**	(metre/second)	
	ft/sec	0.304 8	**m/s**		
	km/hr	$0.277\ 8 \times 10^{-7}$	**m/s**		
	mile/hr	1.609	**km/hr**	(kilometre/hour)	
viscosity	Centipoise	0.001	**Pa•s**	(pascal•second)	
	Centistokes	1.00×10^{-6}	**m^2/s**	(square metre/second)	
volume	fl oz (U.S.)	2.957×10^{-2}	*l*	(litre)	
	quart	0.946	*l*		
	gal (U.S.)	3.785	*l*		
	gal (Imp.)	4.546	*l*		
displacement	in^3	1.639×10^{-2}	***l*		
	**in^3	1.639×10^{-5}	**m^3**		
	**ft^3	2.832×10^{-2}	**m^3**		
	**ft^3	28.32	*l*		
	litre	0.001	**m^3**		
	m^3	1 000	*l*		
			*litre is the preferred unit for engine displacement **solid displacement		
volumetric flow	ft^3/min	28.32	***l*/min**	(litre/minute)	
	ft^3/min	0.472	***l*/s**	(litre/second)	
	ft^3/sec	28.32	*l*/s		
	in^3/min	2.73×10^{-4}	*l*/s		
	oz/hr	2.957×10^{-2}	*l*/hr		
	qt/hr (U.S.)	0.946	*l*/hr		
	qt/hr (Imp.)	1.1365	*l*/hr		
	gal (U.S.)/min	3.785	***l*/min**		
	gal (U.S.)/min	0.063 1	*l*/s		
	gal (U.S.)/hr	3.785	*l*/hr		
	gal (Imp.)/min	4.546	*l*/min		
	gal (Imp.)/min	0.075 8	*l*/s		
	gal (Imp.)/hr	4.546	*l*/hr		
			*from 1–100 ft^3/min, use *l*/min above 100 ft^3/min, use *l*/s		
weight to power	lb/hp	0.608×10^{-3}	kg/W	(kilogram/watt)	
	lb/hp	0.608	**kg/kW**	(kilogram/kilowatt)	

249

Note: Boldface type indicates preferred units.

POSITIONAL TOLERANCE

CONVERSION CHART

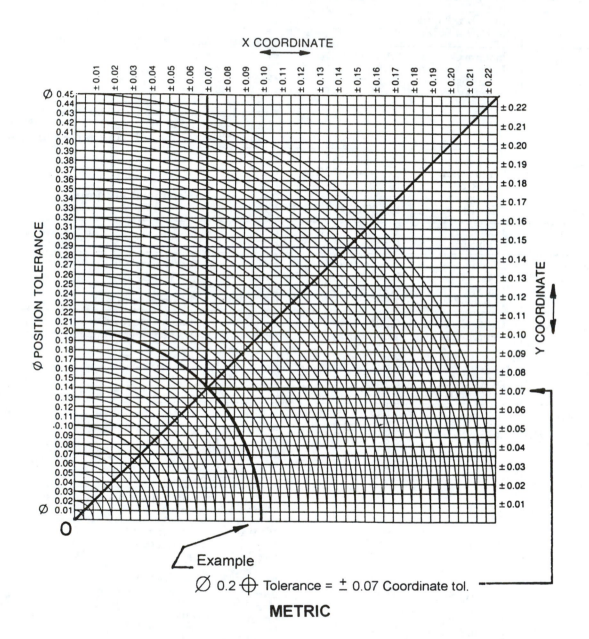

Example

$\emptyset$ 0.2 $\oplus$ Tolerance = $\pm$ 0.07 Coordinate tol.

METRIC

ISO REFERENCE STANDARDS

ISO 1	Standard Reference Temperature for Linear Measurement.
ISO 128	Technical Drawings - General Principles.
ISO 129	Technical Drawings - Definitions and Special Indications.
ISO 286-1	System of Limits and Fits - Basis
ISO 286-2	System of Limits and Fits - Tables.
ISO 406	Tolerancing of Angular and Linear Dimensions.
ISO 1101	Technical Drawings - Geometric Tolerancing.
ISO 1660	Dimensioning and Tolerancing of Profiles.
ISO 1938-1	Inspection of Workpieces - Limit Gages.
ISO 1938-2	Inspection of Workpieces.
ISO 1938-3	Inspection of Workpieces - Guidelines for Inspection.
ISO 2692	Dimensioning and Tolerancing - MMC Principle.
ISO 2692-1	Dimensioning and Tolerancing - LMC Principle.
ISO 3040	Dimensioning and Tolerancing - Cones.
ISO 3274	Instruments for Profile Measurement.
ISO 3650	Gauge Blocks.
ISO 3670	Blanks for Plug and Ring Gauges - General.
ISO 5458	Dimensioning and Tolerancing - Position.
ISO 5459	Dimensioning and Tolerancing - Datums and Datum Systems.
ISO 5460/TR	Tolerances of Form, Orientation, Location and Runout - Verification Principles.
ISO 7083	Symbols for GDT - Proportion and Dimensions.
ISO 8015	Fundamental Tolerancing Principles (Independency).
ISO 8062-2	Castings - GDT.
ISO 10209-1	Technical Product Documentation - Vocabulary.
ISO 10578	Dimensioning and Tolerancing - Projected Tolerance Zone.
ISO 10579	Dimensioning and Tolerancing - Non-Rigid Parts.
ISO 14660-1	Geometric Product Specification - Terms and Definitions.
ISO 14660-2	Geometric Product Specifications - Extracted Terms and Definitions.

Drafts

TC213/WG13N1	Statistical Tolerancing - Terms, Definitions and Symbols.
N2	Statistical Tolerancing - Tolerancing of Mechanical Parts.
N3	Statistical Tolerancing - Actual Values.
N4	Statistical Tolerancing - Assessment of Populations.
N5	Statistical Tolerancing - Analysis of Mechanical Assemblies.

Draft International Standards adopted by the technical committees are approved by ISO member bodies before acceptance as International Standards by the ISO Council. ISO procedures require approval by at least 75% of the member bodies voting. This process allows adoption of standards which may be contrary to a member Country standard, and thus _not_ approved or adopted by that Country.

Y14.5 and ISO COMPARISONS

ISO uses (E) to accomplish the same effect as Rule 1; otherwise,
ISO 8015 is in effect (independency principle), which states that size and
geometric tolerances may exist independently and therefore be additive.

Y14.5 applies *position tolerance controls* to a cylinder, width, or sphere.
ISO may also apply position tolerance to surface, line, or point.

Y14.5 *Composite positional tolerance control*: upper control is location to datum
frame, and lower control is orientation to datum frame and individual feature spacing.
ISO composite positional tolerancing implies two *separate positional* requirements.

ISO uses qualifying notes with *flatness* (not convex or concave). Y14.5 does not show
examples.

Y14.5 applies *flatness* to a single surface only. ISO allows *flatness* to be applied to
multiple coplanar surfaces under certain conditions.

Y14.5 applies *symmetry* to planar features of size whereas ISO applies *symmetry* to
planar or diametral features as well as lines, or in a bidirectional manner.

Y14.5 applies *concentricity* to surface of revolution, RFS only, resulting in *derived
median points*, constituting a *derived median line*. ISO also applies to circular elements,
RFS/MMC or LMC, resulting in axial control (same as *true position* RFS in Y14.5).

Y14.5 covers mathematically defined surface geometry, which could then be used as a
datum surface. ISO does not deal with mathematical surfaces.

The *diameter symbol* precedes all diametral feature values in Y14.5. ISO allows an
omission where shape is obvious. The same is true with the *square symbol*.

Y14.5 requires specific datum sequence in control frame. ISO also allows a nonspecific
ambiguous order if shown without line separations and when sequence is noncritical.

Y14.5 and ISO symbol differences are covered in chapter 1, pages 4 and 5.

Y14.5 employs US standard *third angle projection*, whereas ISO 1101 allows either first
or third angle projection, provided that there is clear notation on the drawing.

DIMENSIONING AND TOLERANCING CHECKLIST ASME Y14.5M-1994

DOES THE DRAWING SPECIFY CORRECT REFERENCE STANDARD: ANSI, ASME, ISO?

DATUMS AND DATUM FRAMES
Is the datum selection and precedence consistent with intended design function and feature relationships?

Are secondary/tertiary datums specified, and are they controlled to the primary datum?

Is the feature control frame constructed correctly? Datum reference frames?

Are the selected datum features accurate, accessible, and of reasonable size to ensure repeatable measurement results?

Where required, is a tertiary (clocking) datum required and/or specified?

Do the basic dimensions for a feature location originate from the same datums referenced in the control frame?

Does the placement of the datum symbol clearly define the datum feature surface, axis, or centerplane?

Have the datum symbols been applied properly per the reference standard named? (The 1994 ASME standard uses the ISO symbology.)

RULES, GENERAL AND SPECIFIC
Are the limits imposed by Rule 1 (LMC local size and MMC envelope) sufficient to control the feature or design, or are other controls (form, orientation, runout, etc.) required?

Is the implied RFS (Rule 2) intended on all feature controls, or does MMC/LMC need to be considered? If so, is the symbol specified?

The PD is implied for all screw thread controls. Is this what you want?

Is it understood that straightness controls must be contained within orientation or position controls?

Do features controlled by basic dimensions have an associated tolerance?

Is it clear, unless stated otherwise, that all dimensions and tolerances apply with the part in the unrestrained (free) state, at 20 degrees C, and for the full feature depth, length, or width?

Do parts and features require constraint to meet design control requirements? If so, are these constraints noted, such as toque specs, loads, and locations?

GENERAL DESIGN ISSUES
Where a cylindrical tolerance zone is intended, has the diameter symbol been included in the feature control frame?

Are coaxial features surfaces controlled to datum features by use of runout, position, or profile? Concentricity controls axial elements, and is the last choice.

Are symmetrically located features controlled? RFS applies.

For controlling radii, are the differences between CR and R understood and specified?

Has the position tolerance been calculated correctly for either a floating or fixed fastener application? T = H - F floating, T = (H - F)/2 fixed.

Is the difference between *composite* and *single segment* position or profile controls understood?

If composite position or profile tolerancing controls are used, is the *orientation principle* in the lower control frame (and datum references) correctly illustrated?

Has MMC been considered to take full advantage of all tolerances, including bonus tolerances as well as full interchangability allowed by functional gaging?

Is minimum wall thickness a design constraint? When tolerancing features at MMC, have the resulting effects of minimum wall thickness been considered (features and datums at LMC with all bonus tolerances in effect)?

DIMENSIONING AND TOLERANCING CHECKLIST (cont.)

Have the implications of *virtual and resultant condition* on features of size been considered and understood? (size tolerance + geometric tolerance + bonus tolerances)

Has *projected tolerance* been considered on fixed fastener feature applications?

SPECIAL CONSIDERATIONS

If *statistical tolerancing* is required or used, have the practices and symbols of ASME Y14.5M-1994 been correctly applied?

If the *boundary principle* is applied to position or profile tolerancing, are the correct methods per Y14.5 applied?

If the *average diameter principle* is applied, the *circularity tolerance* is larger than the size tolerance.

If a drawing revision is being made, is the revision compatible with other mating parts or features?

If a revision per a recent standard is being applied to an older drawing, have the proper standard references been recorded on the drawing?

Is it necessary to carry a design or specification revision to the next drawing level?

GEOMETRIC TOLERANCE MATRIX

TOLERANCE TYPE	CHARACTERISTIC	SYMBOL	CONTROL AT AXIS-MED PLANE OR SURFACE	MODIFIERS FEATURE	MODIFIERS DATUM	TOLERANCE ZONE SHAPE
FORM	STRAIGHTNESS LINE ELEMENT		SURFACE	NO	N/A	PARALLEL LINE ELEMENTS
	STRAIGHTNESS AXIS/ MED PLANE		AXIS/ MED PL	YES	N/A	CYLINDRICAL / PARALLEL PLANES
	FLATNESS		SURFACE	NO	N/A	PARALLEL PLANES
	CIRCULARITY		SURFACE	NO	N/A	SPACE BETWEEN CONCENTRIC CIRCLES
	CYLINDRICITY		SURFACE	NO	N/A	SPACE BETWEEN CONCENTRIC CYLINDERS
ORIENTATION	ANGULARITY		1 A / MP / SUR	YES,IF SIZE FEATURE	YES,IF SIZE FEATURE	1 PARALLEL PLANES / CYLINDERS
	PERPENDICULARITY		1 A / MP / SUR	YES,IF SIZE FEATURE	YES,IF SIZE FEATURE	1 PARALLEL PLANES / CYLINDERS
	PARALLELISM		1 A / MP / SUR	YES,IF SIZE FEATURE	YES,IF SIZE FEATURE	1 PARALLEL PLANES / CYLINDERS
RUNOUT	CIRCULAR RUNOUT		2 SURFACE	NO	NO	SPACE BETWEEN CONCENTRIC DATUM RELATED CIRCULAR ELEMENTS
	TOTAL RUNOUT		2 SURFACE	NO	NO	SPACE BETWEEN CONCENTRIC DATUM RELATED CYLINDERS
PROFILE	3 PROFILE LINE		SURFACE	NO	YES,IF SIZE FEATURE	3 2D PROFILE LINE BOUNDARY
	3 PROFILE SURFACE		SURFACE	NO	YES,IF SIZE FEATURE	3 3D PROFILE SURFACE BOUNDARY
LOCATION	3 POSITION		AXIS / MED PL	YES	YES,IF SIZE FEATURE	CYLINDER / PARALLEL PLANES
	CONCENTRICITY		4 AXIS	NO	NO	CYLINDER
	SYMMETRY		4 MED PL	NO	NO	PARALLEL PLANES

1 WORDS "LINE ELEMENTS" WILL CHANGE TOLERANCE ZONE TO PARALLEL LINES.

2 COMPOSITE CONTROLS OF FORM, LOCATION, AND ORIENTATION.

3 PROFILE AND POSITION MAY ALSO BE USED WITHOUT DATUMS. MAY BE USED IN COMBINATION AS THE "BOUNDARY" CONCEPT.

4 ACTUAL CONTROL OF OPPOSED MEDIAN POINT RATHER THAN AXIS OR PLANE.

RESOURCES AND ADDRESSES

ANSI
1430 Broadway
New York, NY 10018
212-354-3300

SME
One Dearborn Dr.
P.O. Box 930
Dearborn, MI 48121-0930

National Center for Standards and
Certification Information
National Institute for Standards and
Technology
U.S. Dept. of Commerce
Administration Bldg. Room A629
Gaithersburg, MD 20899
301-975-4040

International Standards Organization
(ISO)
Rue de Varembe 1
CH-1211 Geneva 20, Switzerland
(41 22) 34 12 40

American Society for Quality Control
310 W. Wisconsin Ave. Suite 500
Milwaukee, WI 53203
414-272-8575   800-248-1946

ASME Order Dept.
22 Law Dr. Box 2300
Fairfield, NJ 07007-2300
800-843-2763

GDT Textbook orders
800-922-0579

Gary Gooldy
3240 Hillcrest Dr.
Columbus, IN 47203
812-372-9693

GENERAL TESTS

This book concludes with a series of tests. The answers to these tests are available in a separate publication. Good Luck!

General Test 1

2X ⌀.255-.265(holes)

| ⊕ | ⌀.005 Ⓜ | C | B | A |

1. The Maximum Material Condition of size of the hole is:
   A. .250   B. .255   C. .260   D. .265

2. Where the hole size is ⌀.264, the size of the tolerance zone is:
   A. ⌀.005   B. ⌀.010   C. ⌀.012   D. ⌀.014

3. The virtual condition of the hole is:
   A. ⌀.250   B. ⌀.255   C. ⌀.260   D. ⌀.265

4. The shape of the tolerance zone is: A. a width B. a cylinder C. a square.

5. The primary datum reference in the feature control frame is:
   A. datum feature A   B. datum feature B   C. datum feature C.

6. The names of the following symbols are:
   A. ⊕ _____   B. ⊥ _____   C. Ⓛ _____
   D. ⌀ _____   E. ⌂ _____   F. ↗ _____

7. In a feature control frame, geometric controls are applied (at MMC) (at LMC) (at RFS)_____ unless otherwise specified.

8. How many minimum points of contact are needed to establish datum planes?
   A. Primary _____   B. Secondary _____
   C. Tertiary _____

9. A basic dimension: A. establishes a theoretically perfect feature size of location.   B. directs you to the tolerance block. C. establishes a feature MMC.

10. Concentricity is best used as a control for: A. position at MMC
    B. coaxiality   C. interchangeability

11. Concentricity and Runout controls are to apply: A. at LMC   B. at MMC
    C. regardless of Feature Size.   D. any of above

12. RFS applied to a tolerance value means that:
    A. The tolerance zones get smaller as the features get smaller.
    B. The tolerance zones remain the same size at any increment of feature size.
    C. The tolerance zones increase as the feature depart from their MMC size.

13. Tolerance zone values in a feature control frame are understood to be:
    A. plus and minus   B. totals   C. a bonus

257

14. Datums are: A. actual part surfaces   B. Alphabetical features
    C. theoretically exact points axes and planes.

15. Perpendicularity, Angularity, and Parallelism controls govern the:
    A. shape   B. location   C. Orientation between features.

16. Tolerance zones are: A. sometimes   B. always   C. never
    located and oriented relative to datums.

17. If the primary datum is an axis and the secondary datum is a plane, the secondary
    datum may contact the datum reference at
    A. one-point   B. two-points   C. three-points

18. Where a datum feature is a plane surface, measurements from that feature to other
    features of the part are taken from:
    A. the actual surface of the part.   B. the datum simulated from that datum
    feature. C. the theoretical datum surface.

19. The letters LMC stand for: A. Least Manufacturing Cost   B. Least Machining
    Consideration   C. Least Material Condition

20. RULE #1...   A. controls one feature relative to another.   B. controls MMC,
    LMC and RFS.   C. controls perfect form at MMC.   D. controls the
    perpendicularity between plane surfaces.

21. From the drawing below, find minimum distance. **"X"**

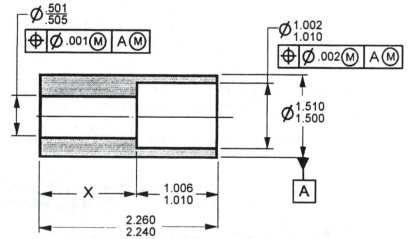

FROM THE DRAWING ABOVE:

22. Using the small hole on the left, what size would the POSITIONAL tolerance
    zone be if the actual hole size is .503?   _____

23. Using the hole on the right, what size would the POSITIONAL tolerance zone
    be if the actual hole size is 1.009?   _____

General Test 2

Indicate in the figure below the pin to be perpendicular.to the bottom surface within 0.25 ∅ MMC.

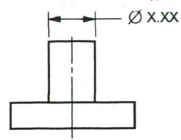

∅ X.XX

Complete the tables below (axial straightness tolerance zones) of the measured sizes given.

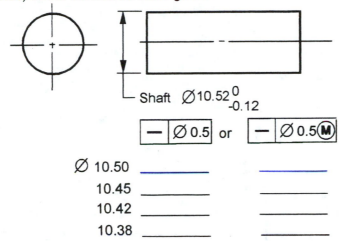

Shaft ∅10.52 $^{0}_{-0.12}$

| — | ∅ 0.5 | or | — | ∅ 0.5 Ⓜ |

∅ 10.50	_____	_____
10.45	_____	_____
10.42	_____	_____
10.38	_____	_____

Complete the tables below (axial perpendicularity tolerance zones) of the measured sizes given.

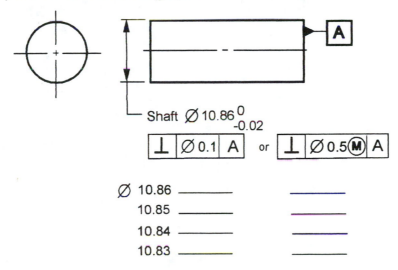

Shaft ∅ 10.86 $^{0}_{-0.02}$

| ⊥ | ∅ 0.1 | A | or | ⊥ | ∅ 0.5 Ⓜ | A |

∅ 10.86	_____	_____
10.85	_____	_____
10.84	_____	_____
10.83	_____	_____

259

In the figure below: regardless of feature size, surface X
must lie between two coaxial boundaries 0.05 apart, having
an angle of 10 degrees with respect to datum feature A.

1. Identify a secondary datum.
2. Tolerance the gage dimension appropriately (+/- 0.03 in
    one direction).
3. Complete the feature control frame.
4. Explain your design logic.

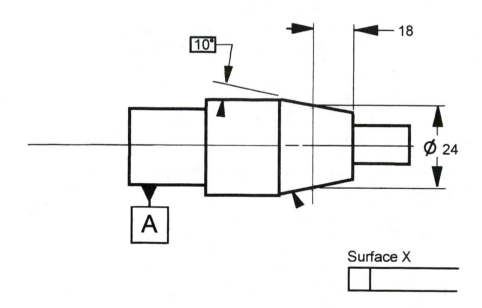

Surface X

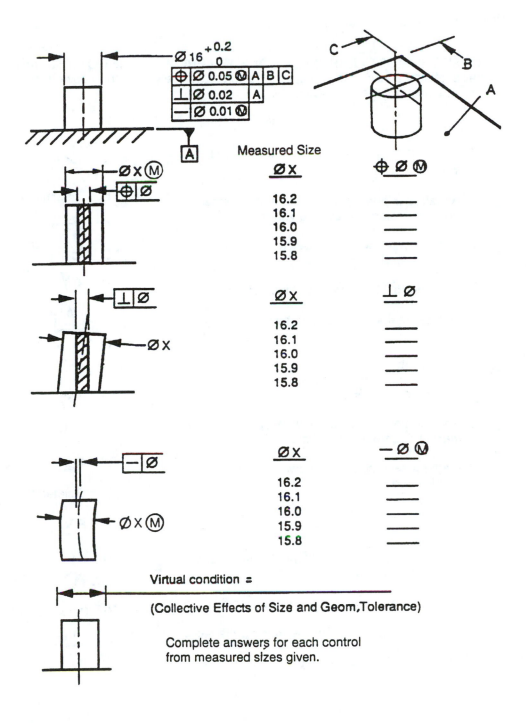

Measured Size

Ø 16 +0.2 / 0

⊕ | Ø 0.05 Ⓜ A B C
⊥ | Ø 0.02 A
— | Ø 0.01 Ⓜ

A

Ø X Ⓜ
⊕ Ø

ØX	⊕ Ø Ⓜ
16.2	
16.1	
16.0	
15.9	
15.8	

⊥ Ø
Ø X

ØX	⊥ Ø
16.2	
16.1	
16.0	
15.9	
15.8	

— Ø
Ø X Ⓜ

ØX	— Ø Ⓜ
16.2	
16.1	
16.0	
15.9	
15.8	

Virtual condition =

(Collective Effects of Size and Geom, Tolerance)

Complete answers for each control
from measured sizes given.

General Test 5

CHOOSE THE MOST APPROPRIATE ANSWER FROM THE LIST ON THE RIGHT.
Not all available answers will be used, and some are used more than once.
Question 7 has multiple answers...

___ 1. Dimensions originating from a datum control the relationship of ....

___ 2. A feature is permitted only the stated tolerance, no bonus tolerance as size varies.

___ 3. A form control with a tolerance zone between two concentric circles.

___ 4. This control identifies a tolerance of location.

___ 5. This geometric control requires a basic angularity dimension.

___ 6. LMC symbol.

___ 7. Implied to apply RFS.

___ 8. Condition of a part feature when it contains the maximum amount of material.

___ 9. A tolerance zone between two parallel planes, parallel to a datum

___ 10. A tolerance zone between two parallel planes, perpendicular to a datum.

___ 11. A geometric form control that allows deviation from a straight line.

___ 12. A tolerance zone between two concentric cylinders.

___ 13. All datum planes intersect at 90 degrees basic.

___ 14. Datum identification symbol.

A. ∠

B. Datum reference frame

C. Form

D. —

E. ⌀

F. [A]

G. .175

H. ⊕

I. Ⓜ

J. ○

K. ◎

L. Related features

M. RFS

N. //

O. Ⓛ

P. Datum target

Q. ⊥

General Test 6

In the figure below, for all features RFS illustrate:

1. Datum surface N to have a total runout of 0.2 FIM relative to datum M.
2. Surface Z:
   Circular runout to be 0.5 relative to primary datum M and secondary datum N.
   Line elements of surface Z to have a 0.3 tolerance from the basic profile between points X and Y.
3. Holes H are to be in true position at MMC relative to secondary datum M (MMC) and primary datum N. The fasteners are 10 mm screws, and the mating part has been assigned 60% of the available tolerance.

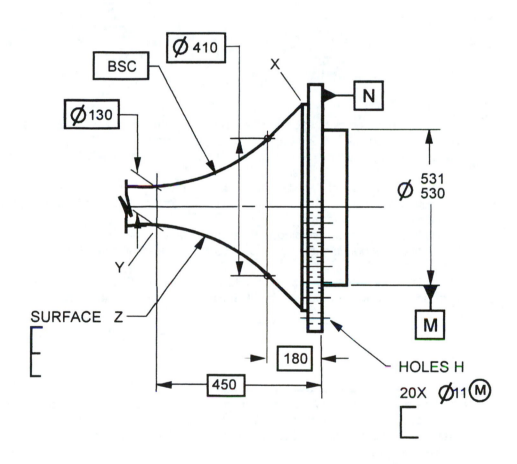

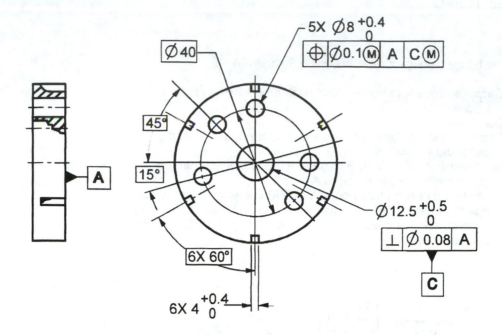

Illustrate: The six grooves to be located from primary datum A and secondary
    datum C (RFS) within 0.2 (MMC), and as a group are datum B.

Illustrate the five small holes also related to tertiary datum B (MMC).

What is the MMC size of the small holes?

What is the MMC virtual condition of the small holes?

Recalculate a positional tolerance for the small holes based on a fixed fastener
    condition, using M6 screws_____.

What is the MMC virtual condition of the large center hole?

What is the total *bonus tolerance* allowed for the five small holes?

In addition to the above, it has been determined that the five small holes must be
    located from datum C within 0.05 RFS. How would you illustrate this?

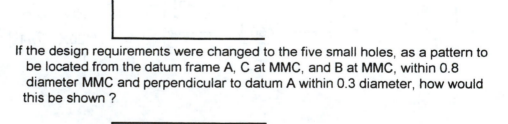

If the design requirements were changed to the five small holes, as a pattern to
    be located from the datum frame A, C at MMC, and B at MMC, within 0.8
    diameter MMC and perpendicular to datum A within 0.3 diameter, how would
    this be shown ?

General Test 8

True or False

1. Rule 1 does not apply to feature relationships.

2. Y14.5 applies position tolerances to a cylindrical feature, a feature width, or a spherical feature.

3. MMC is implied for all feature controls. RFS or LMC must be specified where needed.

4. Y14.5 applies flatness to a single feature surface only.

5. MMC and LMC can only be applied to a feature of size.

6. Straightness controls apply only at feature surfaces and therefore cannot be modified MMC or LMC.

7. Mathematically defined surface geometry can be used as a datum surface.

8. When used in combination with orientation or position controls, the straightness control shall be within the specified orientation or position control tolerance.

9. The diameter symbol follows all diametral feature values.

10. Straightness controls may be applied to feature surface line elements or to feature axes or centerplanes.

11. Datum targets may not be referenced in the feature control frame.

12. Y14.5 employs third angle projection for engineering drawings, whereas ISO 1101 allows either first or third angle projection.

13. Secondary and tertiary datum features of size must be simulated at their LMC Conditions.

14. Centerplanes and axes are never specified as datums.

15. The tangent plane concept differs from parallelism, in that the surface form is not controlled.

16. Profile controls may be applied with or without datums.

17.   Profile controls are not to be used in conjunction with size tolerances.

18.   Profile controls may be used to control coplanar relationships of two
       or more surfaces.

19.   Orientation controls may be applied in combination with position controls,
       invoking the "Boundary" concept.

20.   Profile controls are applied only to features MMC and datums RFS.

21.   Runout controls are composite surface controls and contain all
       errors of size, form, orientation, and location.

22.   All coaxial features shall be controlled by position, runout, profile and
       concentricity. Consider concentricity first.

23.   Where a cylindrical tolerance zone is intended, a diameter symbol
       is understood and implied.

24.   Profile controls may be used to control irregular shapes.

25.   Zero tolerancing offers maximum manufacturing flexibility and
       is recommended under most circumstances.

26.   With MMC, bonus tolerances may have an adverse affect on wall
       thickness,

27.   Virtual/resultant conditions for size features include the combined
       extremes of size and geometric tolerances.

28.   Basic dimensions do not require a specified associated tolerance.

29.   Flatness also controls straightness of surface line elements.

30.   Profile controls may be applied as composite controls.

31.   The tolerance zone for circularity is described as a diameter zone.

32.   A cylindricity tolerance zone is described as two concentric circles.

33.   Concentricity is used to control the median line of a cylindrical feature
       relative to a datum axis.

34.   A profile tolerance zone is implied as bilateral from the basic profile.

35.   Runout or position controls are generally preferred over concentricity.

36.   A coaxial relationship of a cylindrical feature to a datum axis may be accomplished using a position control RFS.

37.   LMC is not appropriate when minimum wall thickness is a concern.

38.   In a feature control frame, datum references must be in alphabetical order.

39.   Datum features are considered theoretically exact.

40.   To control Parallelism of surface line elements to a datum, the words EACH ELEMENT or LINE ELEMENTS must be added beneath the feature control frame.

41.   Orientation control tolerances are contained within size tolerance.

Coaxial Controls

From the descriptions below, select the correct
descrition and place beside each control frame.

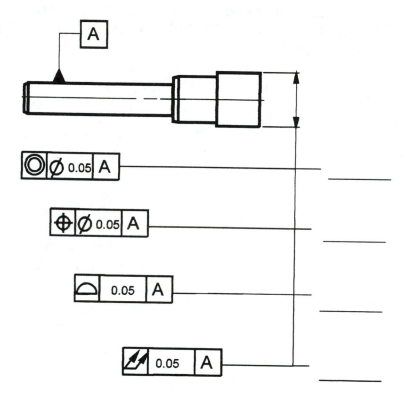

A. Controls feature surface to datum axis

B. Controls feature surface, size, form, locatlon,
and/or orientation to datum axis

C. Controls feature median points to datum axis

D. Controls controls feature axis to datum axis

EDGE DISTANCE EXERCISE

From the figure below, determine the *minimum* available wall thickness for dimensions A and B. Assume that a perfectly rectangular part has been produced.

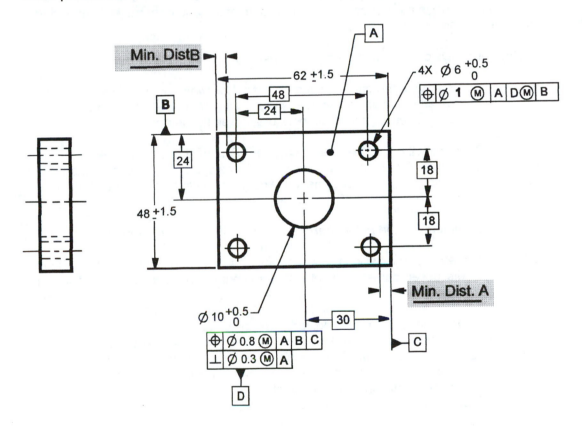

General Test 11 Functional Analysis

Interfacing and mating parts and features have a priority of mating contact based on functional significance as well as fits and clearances. Determining this priority is the process of determining the primary, secondary, and tertiary datums and datum framework. The primary datum exhibits the most important contact, gives the greatest degree of control, and transfers most critical operational design influences. Because parts cannot be perfectly produced, any manufacturing or measurement error will be transmitted to the secondary and tertiary datums. The secondary datum offers part stabilization, while the tertiary datum stops all rotation, and generally "locks" the part in the datum framework.

A typical analysis process may be as follows:

1. Determine the part or assembly *critical function*, that is, transfer dynamic operation functions and influences (rotary or linear motion, torque, bending, etc.): mount, support, or align static fits and surfaces.

2. Identify the part features associated with the above and *develop the priority* of these issues. For example, in a dynamic, or bearing/shaft application, is the alignment of axes or the seating against a mounting face more important?

3. Determine the *datum framework*: primary, secondary, tertiary.

4. Develop and assign the size tolerances based on *preferred limits and fits* discussed earlier, geometric feature controls along with special notes or concerns.

On the following page is a partial design of a sliding shaft used as an actuating switch device. In the design are some critical moving fits, static alignment fits and clearance fits. Analyze this design. You may have to improvise somewhat, but give the analysis process a try. See what you come up with. You may find the diagram and process on page 244 helpful.
   There is usually more than one way!

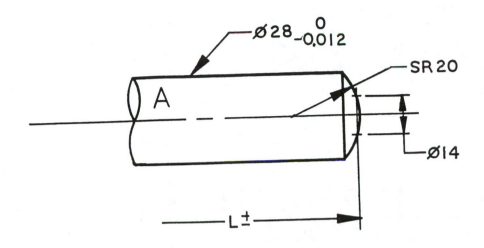

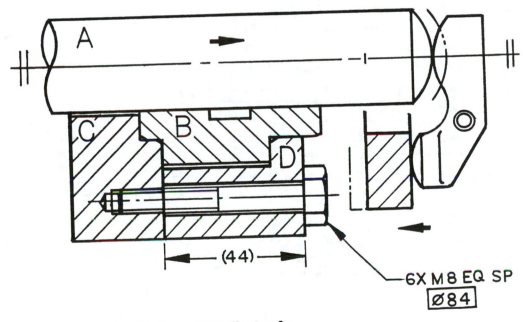

Rules Of Thumb

• Form Tolerance - (1/2 Size Tolerance) MAX

271

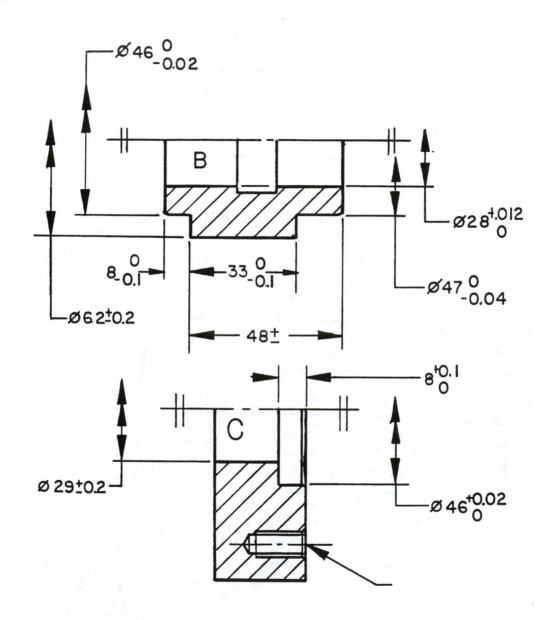

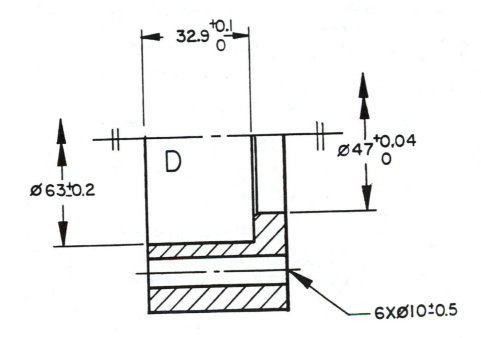

GLOSSARY

Actual Local Size - measured size at any cross section of a feature.

Actual Mating Size - value of minimum/maximum mating envelope.

Actual Size - measured size; includes local and mating size.

Angularity - surface, axis, or centerplane at any angle (other than 90 degrees) relative to a datum.

ANSI - American National Standards Institute.

ASME - American Society of Mechanical Engineers.

Basic - a numerical value that describes an exact theoretical size, shape, location, etc. of a feature, datum, or target. The value is placed within a box or rectangle.

Basic Size - the size from which limits and tolerances are derived.

Bilateral Tolerance - a tolerance that exists in two directions simultaneously from a given dimension.

Bonus Tolerance - a tolerance that may be added due to size variation.

Boundary, Inner - smallest possible feature envelope (cylinder) for an internal feature at MMC or external feature at LMC.

Boundary, Outer - largest possible feature envelope (cylinder) for an external feature at MMC or internal feature LMC.

Calipers - tool to make opposed or cross sectional measurements.

Centerline - middle or median line of a feature.

Centerplane - middle or median plane of a feature.

Chamfer - edge break or cut of a feature shoulder to remove sharp edges.

Circularity - form control for the surface elements (at a cross section) of a feature of revolution (cylinder, cone, sphere). Also called *roundness*.

Circular Runout - composite surface control at the cross section of a feature relative to a datum axis.

Clearance Fit - condition where the internal part size limit is always smaller than the external mating size limit.

Coaxiality - condition of two or more features sharing the same axis.

Composite Tolerance - tolerance providing a relationship to a datum framework as well as feature relationship (i.e., profile/position).

Concentricity - condition where two or more features of various shapes are in line with a datum feature axis, both/all at RFS.

Controlled Radius - radius without flats or reversals, and within the size limits.

Contour - see Profile Tolerance.

Coordinate Measuring Machine (CMM) - device that measures, records, calculates, and analyzes measured feature variations.

Coplanarity - two or more surfaces that share the same plane.

Counterbore - stepped increase in feature hole diameter.

Cylindricity - form control with all surface elements equidistant from a common axis.

Datum - theoretically exact point, line (axis), or plane.

Datum Axis - axis established by the datum feature.

Datum Feature - actual physical feature (surface) of a part.

Datum Feature of Size - physical feature with size variation.

Datum Feature Symbol - reference letter contained within a box.

Datum Plane - theoretically exact plane.

Datum Reference Frame(work) - three mutually perpendicular planes.

Datum Simulator - processing or inspection equipment surfaces.

Datum Surface - actual feature surface (flat surface, hole, slot, etc.) used to establish the theoretically perfect datum.

Datum Target - points, lines, or areas used for consistency and repeatability in establishing datum surfaces.

Datum Target Symbol - large circle containing letters and numbers for for identification of targets.

Diameter Symbol - circle with diagonal line.

Dimension - numerical value - used with lines, notes, symbols, etc. to define part characteristics.

Dial Indicator - device to measure variation from any desired condition.

Envelope, Actual Mating:

   External - smallest size that can be circumscribed about the feature so that it contacts the high points of the feature surface.

   Internal - largest size that can be inscribed within the feature so that it contacts the high points of the feature surface.

Extension Line - line to extend dimensional features.

Feature - general term for a physical portion of a part (hole, surface, slot, thread, etc.).

Feature Axis - the axis of the features true geometric counterpart.

Feature Centerplane - the centerplane of the features true geometric counterpart.

Feature Control Frame - rectangular boxes that form specification controls in symbol sentences.

Feature of Size - feature that when changed, will affect the physical weight of a part, such as; length, width, thickness, hole.

Fit - general term to identify looseness or tightness of assembly conditions.

Fixed Fastener Condition - assembly condition of two or more parts with one having features that are threaded, press fit, or line fit in nature, with the fastener or mating feature involved.

Flatness - form control for a surface with all surface elements in one plane.

Floating Fastener Condition - assembly condition of two or more parts with assembly clearance in both parts.

Force Fit - see Interference Fit.

Form Tolerance - category of tolerance controls of individual features such as flatness, circularity, straightness, or cylindricity.

Free State Variation - condition of a part, that when unrestrained (free) the dimensional limits may change, or are unstable.

Full Indicator Movement - difference between minimum and maximum limits read on and indicator device.

Functional Gage - a receiver gage that will accept the part with no force applied.

Geometric Characteristics - basic elements used for specification and control in *geometric tolerancing* such as form, orientation, profile, runout, and location controls.

Geometric Tolerance - the application of geometric characteristic controls.

Implied Datum - unspecified datum implied by the dimensioning structure of the engineering drawing.

Interference Fit - condition where the size limits of two parts will always result in an interference or force fit.

Irregular Curve - curve with no constant radius.

Keyway/Keyslot - slot to control a fixed relationship (rotational) between parts.

Least Material Condition - feature of size contains the least material or when the part weighs the least.

Limit Dimension - both minimum and maximum dimensions are specified.

Limits of Size - specified maximum and minimum size.

Line Fit - condition with size limits that result in a zero clearance between parts.

275

Location Tolerance - specification allowing departure from an exact location (includes position, concentricity, and symmetry tolerance).

Material Condition - MMC, RFS or LMC.

Maximum Dimension - the extreme upper limit dimension.

Maximum Material Condition (MMC) - condition where a feature of size contains the most material and weighs the most.

Median Line - line derived from the centerpoints of cross sectional measurements of a size feature.

Median Plane - plane derived from the centerlines of cross-sectional measurements of a size feature.

Micrometer - measuring device for taking opposed diametral measurements.

Minimum Dimension - the extreme lower limit dimension.

Modifier - term or symbol to describe appropriate material condition applications (RFS, MMC, LMC).

Multiple Datum Reference Frames - condition where features are controlled to more than one set of datum references.

Nominal Size - term for general identification.

Orientation Tolerance - category of tolerances that control one feature relationship to another (angularity, parallelism, and perpendicularity).

Origin of Measurement - points, lines, or planes from which measurements are taken.

Paper Gaging - graphical and mathematical analysis of inspection data.

Parallelism - condition of a feature surface, axis or line element equidistant from a datum plane or axis.

Perfect Form at MMC - extreme size feature envelope.

Perpendicularity - condition of a feature surface, axis, or line element at a right angle to a datum plane or axis.

Position Tolerance - tolerance zone within which a feature axis or centerplane is allowed to vary.

Primary Datum - datum with the greatest influence on a given set of design criteria. The first datum listed in the control frame.

Profile, All Around - method for applying constant perimeter control.

Profile Tolerance - condition that allows a surface or line element to vary uniformly from a desired true profile.

Projected Tolerance Zone - tolerance zone that is projected above or below the feature surface, normally equal to the thickness of the mating part.

Reference Dimension - dimension without tolerance, enclosed by parentheses, for information only, not intended for application by manufacturing or inspection.

Regardless of Feature Size (RFS) - condition that indicates that a feature tolerance applies regardless of the feature size, within the size limits.

Resultant Condition - a variable boundary generated by the collective effects of size tolerances, all bonus tolerances, geometric tolerances and is generally considered the opposite of virtual condition.

Roundness - see Circularity.

Rules - preconditions, givens.

Runout - see Circular Runout and Total Runout.

Runout Tolerance - composite or total value measured at a feature surface, relative to a datum axis through one full revolution of the part.

Secondary Datum - datum with lesser influence on a given set of design criteria. The second datum listed in the control frame.

Simulated Datum - datum simulated by inspection or processing equipment surfaces, or datum simulators.

Simultaneous Datum Features - two features that when used together, make a single datum reference plane or axis.

Size Dimension - values that define a part or feature size.

Size Feature - see Feature of Size.

Size Tolerance - states allowed departure from desired size.

Spherical Feature - a feature with all points on the surface equidistant from a given point of origin.

Squareness - see Perpendicularity.

Statistical Tolerance - the mathematical manipulation of data.

Straightness - form control where line elements of a feature surface or a feature axis are a straight line.

Surface Texture - primary (roughness) and secondary (waviness) surface tool marks generated by processing equipment. Also includes the lay of tool marks.

Symmetry - condition of a feature equally disposed about the centerplane of a datum feature.

Tangent Plane - the plane that contacts the high points of a feature surface.

Taper - diametral change per unit length of a conical feature.

Target - see Datum Target.

Tertiary Datum - datum with the least influence on a given set of design criteria. The third datum listed in the control frame.

Times (Places) - the number of occurrences of a feature or dimension, indicated by an "X."

Tolerance - total permissible variation of a specification.

Total Runout - composite control of all surface elements of a feature surface relative to a datum axis.

Transition Fit - condition where the size limits of two parts may result in either a clearance or interference.

True Geometric Counterpart - theoretically perfect feature, or datum feature virtual condition boundary. Also actual mating envelope.

True Position - theoretically exact location relative to a datum reference or other feature.

Unilateral Tolerance - tolerance that exists in one direction from a specified dimension.

Virtual Condition - condition created by combined effects of size tolerance (MMC or LMC) and any geometric tolerance.

Waviness - see Surface Texture.

Workpiece - a part or assembly in process of manufacture or evaluation.

Zero Tolerancing - tolerancing technique in which tolerance is allowed based on a features departure from MMC or LMC size limits only.

INDEX

278